Synthese Library

Studies in Epistemology, Logic, Methodology, and Philosophy of Science

Volume 414

The aim of *Synthese Library* is to provide a forum for the best current work in the methodology and philosophy of science and in epistemology. A wide variety of different approaches have traditionally been represented in the Library, and every effort is made to maintain this variety, not for its own sake, but because we believe that there are many fruitful and illuminating approaches to the philosophy of science and related disciplines.

Special attention is paid to methodological studies which illustrate the interplay of empirical and philosophical viewpoints and to contributions to the formal (logical, set-theoretical, mathematical, information-theoretical, decision-theoretical, etc.) methodology of empirical sciences. Likewise, the applications of logical methods to epistemology as well as philosophically and methodologically relevant studies in logic are strongly encouraged. The emphasis on logic will be tempered by interest in the psychological, historical, and sociological aspects of science.

Besides monographs *Synthese Library* publishes thematically unified anthologies and edited volumes with a well-defined topical focus inside the aim and scope of the book series. The contributions in the volumes are expected to be focused and structurally organized in accordance with the central theme(s), and should be tied together by an extensive editorial introduction or set of introductions if the volume is divided into parts. An extensive bibliography and index are mandatory.

More information about this series at http://www.springer.com/series/6607

J. Acacio de Barros • Carlos Montemayor
Editors

Quanta and Mind

Essays on the Connection between Quantum Mechanics and Consciousness

Springer

Editors
J. Acacio de Barros
School of Humanities and Liberal Studies
San Francisco State University
San Francisco, CA, USA

Carlos Montemayor
Department of Philosophy
San Francisco State University
San Francisco, CA, USA

Synthese Library
ISBN 978-3-030-21907-9 ISBN 978-3-030-21908-6 (eBook)
https://doi.org/10.1007/978-3-030-21908-6

This Springer imprint is published by the registered company Springer Nature Switzerland AG.
The registered company address is: Gewerbestrasse 11, 6330 Cham, Switzerland

Foreword

From the time of Newton, and as had been emphasized by Laplace and others, a distinctive picture of the detailed behaviour of our universe had been presented, whereby its evolution in time would be precisely determined by specific mathematical evolution equations. In view of the remarkable precision according to which the motions of bodies, both celestial and terrestrial, had seemed to be in accordance with these equations, it had seemed highly plausible that this picture might well exactly describe the detailed behaviour of the actual universe in which we live. In this picture, the view was that each material body would be composed of a vast number of various specific types of particles acting upon one another by particular forces, and although the exact detailed form of all the different forces between those constituent particles might not be yet known, this would not affect the overall picture of a deterministic mathematically controlled precise representation of the behaviour of things.

Yet, this picture had always presented an awkwardness in relation to the actions and experiences of conscious human beings, as it seemed to leave no room for an explanation of the impressions that we have, of exerting conscious control over the motions of our bodies, and thence of other physical objects under our bodies' direct influence. Of course, as many have argued, this impression of a freedom in this conscious control—a notion of *free will*—might be some kind of illusion, and the decisions that we appear to be freely making might actually be completely predetermined, as such a viewpoint of the actions of the universe governed completely by Newtonian laws would seem to imply. Some have argued that such Newtonian systems are frequently "chaotic", whereby there is an extreme sensitivity in the dependence of system on the actual initial conditions, but this does not affect the issue of determinism, and all future behaviour is still exactly determined by initial conditions, which may be taken to have been set in the remote past.

This Newtonian particulate picture was amended to some degree in the mid-nineteenth century by the Faraday-Maxwell introduction of an independent electromagnetic field, according to which the Newtonian picture of a world consisting solely of discrete particles had to be modified, and the presence of degrees of freedom in an independent continuous field had to be allowed for. In principle, however, this made little difference to the overall picture (once certain delicate issues concerning individual point-like particles are appropriately sorted out) since, again, one appeared to have some kind of deterministic evolution from initial conditions in the remote past. Einstein's introduction of special relativity in the early twentieth century did not much affect this issue, although the notion of an evolving time had to be looked at in a different way nor, even, did his introduction of general relativity in 1915, provided that we restrict attention to what are called globally hyperbolic space-times, in which the evolution of the entire space-time can be considered to have evolved deterministically from an initial space-like hypersurface. The mathematical description of the universe and all its contents are considered to be laid down in this deterministic way, and there appears to be no room for any kind of independent objective action for a conscious influence. Yet, again, we can take the view that such an impression of "free will" is just an "illusion" of some kind, whereby all of one's actions are predetermined, and consciousness experience is just an *epiphenomenon*, i.e. just "going along for the ride" and not having any actual influence on the behaviour of things!

However, with the introduction of quantum mechanics, the situation appeared to become fundamentally different, since an important element of indeterminism has now become part of the theory. Yet, this indeterminacy itself enters into the theory in a very strange way, which is highly elusive and deeply controversial. There is still a deterministic revolution in the formalism, namely, that of the *quantum state*, which is to evolve precisely according to the clearly defined Schrödinger equation. At this point there is much controversy about the ontological status of the quantum state. Does it *really* express the reality of the physical world, or is it (as the standard Copenhagen interpretation would seem to imply) "all in the mind of the experimenter" simply representing some kind of "maximum knowledge" about the system that allows probabilities to be calculated concerning the result of an experiment that is about to be performed? After the experiment has actually been performed, the policy is to "collapse" the quantum state into one of the alternatives allowed by the experiment, and the universe seems to have made a "choice" of one of these alternatives, instead of following the deterministic Schrödinger evolution.

Is this some sort of objective effect of the experimenter having "consciously observed" the result of the experiment (perhaps in accordance with a viewpoint explicitly put forward by Wigner), thereby defying the Schrödinger evolution of the state? Or are we to preserve the Schrödinger equation at all cost and resign ourselves to some kind of Everett-type "many-worlds" viewpoint according to which the superposition of possible outcomes extends also to the experimenter, who now is taken to be in a superposition of mutually inconsistent experiences? The view is then taken that the experimenter's consciousness is somehow able to bifurcate into separate branches in a way that each branch experiences but a single well-defined outcome to the experiment.

We appear to have four different possibilities concerning the relation between conscious experience and the collapse of the quantum state. First, perhaps they have nothing to do with one another, and we must seek some purely physical explanation, in which the Schrödinger evolution is violated for some reason quite unconnected with the issue of conscious experience. This is a reasonable point of view to take, but this volume explores the more exciting possibility that there might instead be some fundamental connection. In Part I, the point of view is explored that the formalism of quantum mechanics as we currently understand it may have something deep to tell us about the actual nature of conscious experience. This is a proposal for which I have some sympathy, though I take the view that the rules of quantum theory themselves will first need serious modification. In Part II, it is the reverse possibility (perhaps like the Wigner view) that it is the presence of the consciousness phenomenon that will lead us to a better understanding of how the quantum world is actually to be described. In Part III, the issue is explored of what is to me the most exciting possibility that the phenomena of consciousness and of quantum-state collapse may each have something deep to say about the other. These are important and difficult questions, and this book provides numerous original insights towards their resolution.

Emeritus Rouse Ball Professor of Mathematics Roger Penrose
University of Oxford
Oxford, England

Preface

The debate surrounding the interpretations of quantum theory has been lasting for longer than 100 years, with dramatic exchanges among the founders of quantum mechanics as well as the leading physicists of the last century. Although there has been some progress in terms of no-go theorems and experiments, the status of the debate is still quite open, with no clear consensus or even a majority position (Schlosshauer et al. 2013). There are more interpretations that one should desire, considering that these are wildly different approaches to the foundational theory of physics. Furthermore, the available interpretations leave us with a difficult choice: either think of quantum theory as merely epistemic or accept wild ontological premises. If things keep moving in the direction of more, rather than less, interpretations of quantum mechanics, there will be legitimate concerns that the debate is regressing rather than progressing.

Crudely speaking, two main paths have been taken with respect to interpretations. In one direction, most famously espoused by Bohr, we have epistemic interpretations, such as QBism (Fuchs et al. 2014), the Copenhagen interpretation (see Jaeger 2009), or the modal interpretation (van Fraassen 1991). In the epistemic approach, quantum theory tells us nothing about the world but only about what we can say about the world. In the other direction, more in line with the one preferred by Einstein, we have ontic interpretations, such as the many-worlds (Everett 1957; De Witt 1970), the consciousness causes collapse (Stapp 1999), or Bohm's pilot wave (Bohm 1952; Holland 1995). Those interpretations attempt to understand what quantum theory is telling us about what the world actually is.

It is probable that the main reason for a lack of consensus among physicists with respect to interpretations of quantum theory is that physicists are left with very hard choices. On the one hand, the choice is to "give up" on the idea that physics is a way to understand nature, to uncover what the world is made of, and to surrender. According to the epistemic view, we will never understand the microscopic world, and we might as well just surrender. On the other hand, the existing ontic interpretations bring a lot of metaphysical difficulties. For example, the idea that the world splits into (perhaps infinite) many worlds at every interaction of a quantum system with a measurement apparatus seem ludicrous to some. The relatively popular pilot-wave interpretation is plagued with issues of causality (due to its superluminal potential) and seems to indicate that the actual world is not 3+1 dimensional but instead has infinite dimensions. Similarly, the possibility of an almost Cartesian dual-mind/matter ontology makes other physicists cringe.

So, perhaps a way of making progress is to carefully determine the commitments of the interpretations of quantum mechanics. One such commitment is related to the role of the observer. What is considered the Copenhagen interpretation of QM, perhaps the interpretation espoused by the plurality of working physicists (not working on foundations of quantum theory), puts the observer at a central place in the theory, particularly with respect to the involvement of observers that make measurements. Thus, an important goal of the present volume is to bring as much clarity as possible to themes concerning observation and measurement in quantum mechanics, and as neutrally as possible.

There have been about 30 years of debate concerning the nature of phenomenal consciousness. Some interpretations of quantum mechanics that predate the debate on phenomenal consciousness assumed that consciousness was fundamentally related to the formulation of the laws of quantum mechanics—the very laws governing quantum mechanics necessitated consciousness to explain how measurements had an irreversible impact in determining outcomes. But since it was not until recently that we benefited from more precise definitions of consciousness, such as the distinction between access and phenomenal consciousness, traditional interpretations of quantum mechanics that appealed to consciousness may need to be reinterpreted and clarified. This is another central theme of this volume, with a focus on the type of experimental approaches certain interpretations of quanta and mind might entail.

Accordingly, we organized this volume the following way. In Part I, we included articles that draw from quantum theory, and its interpretations, to draw conclusions about the mind and mind-matter interactions. In Part II, articles investigate how different conceptions of the observer's mind and consciousness may help with our understanding of quantum theory and the role of the observer. Finally, in Part III, we find articles that explore how both quantum theory and theories of consciousness may inform our views about the physical (or metaphysical) world.

Unlike some volumes and monographs on the possible interpretations of quantum mechanics, this volume is not meant to emphasize or highlight the spookiness or strangeness of quanta, or the vast enigma of consciousness, or how to deepen the mysteriousness of quanta by combining it with the mystery of consciousness. There is obvious value in identifying what makes these problems very hard, but that has been done fairly successfully in the extant literature. By contrast, the focus of this volume is to provide a neutral and cross-disciplinary perspective on the current understanding of quanta and mind, not in order to favour scepticism or a specific interpretation but rather to provide more clarity to the debate.

Similarly, unlike many contemporary approaches to quanta and mind, the main goal of the volume is not to provide a single perspective that ultimately relates the quantum to the phenomenal. Rather, the main purpose is to provide constructive engagement with many approaches. The issue of clarity not only concerns the sense in which we should define terms like consciousness and mind but also how to improve our understanding of some of the key claims of the interpretations of quantum mechanics, particularly concerning the measurement problem.

This volume grew out of a conference we organized at our university in April 2018: the International Conference on Quanta and Mind. The conference drew researchers from diverse academic areas, such as physics, neurosciences, and philosophy, as well as different geographic regions. We would like to thank Professors Harald Atmanspacher, Paavo Pylkkanen, and Paul Skokowski, who were part of the scientific committee and without whom this conference and therefore this volume would not have happened. We would also like to thank all the participants of the conference who, through their questions and comments, improved the arguments presented in some of the papers. Not all authors were able to come to San Francisco, and we are grateful for their contributions. Finally, we would like to thank Otávio Bueno, editor of Synthese Library, for his comments and support during this project.

San Francisco, CA, USA — J. Acacio de Barros
January 29, 2019 — Carlos Montemayor

References

Bohm, D. (1952). A suggested interpretation of the quantum theory in terms of "Hidden" Variables. I. *Physical Review*, *85*(2), 166–179.
De Witt, B. (1970). Quantum mechanics and reality. *Physics Today*, *23*(9): 30–35
Everett, H. (1957). "Relative State" formulation of quantum mechanics. *Reviews of Modern Physics*, *29*(3), 454–462. https://doi.org/10.1103/RevModPhys.29.454.
Fuchs, C. A., Mermin, N. D., & Schack, R. (2014). An introduction to QBism with an application to the locality of quantum mechanics. *American Journal of Physics*, *82*(8), 749–754.
Holland, P. R. (1995). *The quantum theory of motion: An account of the de Broglie-Bohm causal interpretation of quantum mechanics*. Cambridge: Cambridge University Press.
Jaeger, G. (2009). *Entanglement, information, and the interpretation of quantum mechanics*. Berlin: Springer.
Schlosshauer, M., Kofler, J., & Zeilinger, A. (2013). A snapshot of foundational attitudes toward quantum mechanics. *Studies in History and Philosophy of Science Part B: Studies in History and Philosophy of Modern Physics*, *44*(3), 222–230.
Stapp, H. P. (1999). Attention, intention, and will in quantum physics. *Journal of Consciousness Studies*, *6*(8/9): 143–164.
Van Fraassen, B. C. (1991). *Quantum mechanics: An empiricist view*. Oxford: Oxford University Press.

Contents

Part II Mind Informs Quanta

Part III Quanta and Mind Informs Worldviews

Part I
Quanta Informs Mind

Chapter 1
Free Will in a Quantum World?

Valia Allori

Abstract In this paper, I argue that Conway and Kochen's Free Will Theorem (Conway and Kochen 2006, 2009) to the conclusion that quantum mechanics and relativity entail freedom for the particles, does not change the situation in favor of a libertarian position as they would like. In fact, the theorem more or less implicitly assumes that people are free, and thus it begs the question. Moreover, it does not prove neither that if people are free, so are particles, nor that the property people possess when they are said to be free is the same as the one particles possess when they are claimed to be free. I then analyze the Free State Theorem (Conway and Kochen 2009), which generalizes the Free Will Theorem without the assumption that people are free, and I show that it does not prove anything about free will, since the notion of freedom for particles is either inconsistent, or it does not concern our common understanding of freedom. In both cases, the Free Will Theorem and the Free State Theorem do not provide any enlightenment on the constraints physics can pose on free will.

Keywords Free will theorem · Strong free will theorem · Free state theorem · Nonlocality · Compatibilist free will · Libertarian free will · Free will · Quantum mechanics

---

V. Allori (✉)
Department of Philosophy, Northern Illinois University, Dekalb, IL, USA
e-mail: vallori@niu.edu

J. Acacio de Barros, C. Montemayor (eds.), *Quanta and Mind*, Synthese Library 414, https://doi.org/10.1007/978-3-030-21908-6_1

1.1 Introduction

The debate over free will and the development of physics are intertwined. Newtonian mechanics, the prototypical deterministic theory,[1] is in tension with free will: the laws of nature control us as the puppet master controls the puppet. Some have resorted to the indeterminism of quantum mechanics. However, free will is different from randomness, so it is also incompatible with indeterminism.[2] Nevertheless, recently John Conway and Simon Kochen (2006, 2009) have proven a theorem, which they call the Free Will Theorem, to the conclusion that quantum mechanics, no matter whether deterministic or stochastic, together with relativity not only are compatible with free will but, in some sense, also entails it.[3]

The theorem has received a lot of attention by the physical community and by a more popular audience,[4] but has not been discussed much within the philosophical community, even if Conway and Kochen themselves encourage the philosophers to have a look to their result.[5] Others have previously attempted the similar task of connecting free will with quantum mechanics[6] and people have already responded to them.[7] However, this time the result looks less speculative: a theorem, if sound, is more than compelling evidence of the existence of a libertarian free will. Moreover this theorem, if sound, follows directly from the quantum formalism and from relativity theory, without further speculations. As such, it seems to have broader implications than, for instance, Henry Stapp's theory of free will, which relies on a particular interpretation of quantum mechanics in which the mind is collapsing the wave function (Stapp 1993, 1995, 2017). Anyway, there may be various reasons why philosophers may not have engaged much with this result. One could be that often arguments from quantum mechanics are inconclusive, given the controversial nature of the theory.[8] Another, that the paper is too technical. Be that as it may, in

[1]Some have questioned the extent to which Newtonian mechanics is deterministic. See, for instance, (Earman 1986; Norton 2008). However, for the purpose of this paper we can ignore these subtleties since we are concerned with quantum mechanics.

[2]See, for instance, (Searle 1984; Strawson 1986; Pinker 1997; Clarke 2003; Balaguer 2004; Kane 1996). The basic idea is that laws, deterministic or stochastic, are still 'in charge' of future actions, we never are: if we are string puppets, the fact that sometimes the strings may jerk randomly does not change the fact that we do not decides how we move.

[3]A more precise statement of Conway and Kochen's thesis will be made clear later in the paper.

[4]*The New Scientist* (Merali 2006) has also reported it.

[5]When discussing some features of the free will compatible with quantum mechanics, they write that their remarks "might also interest some philosophers of free will" (Conway and Kochen 2006), p. 1465).

[6]See, most notably (Kane 1996; Compton 1935; Popper 1972; Nozick 1981; van Inwagen 1983; Penrose 1994; O'Connor 1995; Stapp 1991).

[7]See for instance (Loewer 2003).

[8]To give an example of this attitude, even if quantum nonlocaltiy seemed to provide a knock down argument against Humean supervenience, David Lewis wrote: "if physics tells me that it is false, I wouldn't grieve [...] But I am not ready to take lessons in ontology form quantum physics as it

this paper I aim to analyze what the theorem assumes, what it concludes and whether it sheds a new light on the free will debate.

Here is a map of the paper. In Sect. 1.2 I present the assumptions and the structure of the proof to the conclusion that quantum mechanics and relativity entail free will for particles. In Sect. 1.3 I present the objections raised against the Free Will Theorem present in the physical literature, essentially that one of the assumption, namely MIN, is false. Instead, in Sect. 1.4 I focus on the impact of the Free Will Theorem on the free will debate. First, I observe that the theorem is question begging, since freedom is assumed in the proof. Then I argue that the authors do not even prove that if people are free, then so are particles. Moreover, I argue that if the criticism in Sect. 1.3 are sound, then the theorem actually disprove locality rather than proving freedom of the will (Sect. 1.4.2). In addition, even if such criticisms are incorrect, the meaning of "free will" used for people is not necessarily the same as the one used for particles, and in general it seems absurd to think that the meanings are the same (Sect. 1.4.3). Finally, I present Conway and Kochen's Free State Theorem, which generalizes the Free Will Theorem without the assumption that people are free. I show that the problem with this theorem is that the notion of freedom for particles is either in tension with the assumptions made by Conway and Kochen, or it is a species of randomness, and therefore not freedom (Sect. 1.4.4). Therefore, I conclude that the Free Will Theorem, in all its varieties, is a nice piece of mathematical work but its name suggests much more than it actually does it does not entail anything about freedom, neither for us, nor for the particles.

1.2 The Free Will Theorem: SPIN, TWIN, FIN, MIN and DET

In their proof Conway and Kochen consider a particular experimental situation and assume few axioms called SPIN, TWIN, FIN (later, MIN), and DET. The experiment involves a pair of particles, a and b, with total spin 1 which are traveling in opposite directions, and two experimenters, A and B, that can perform experiments on the spin of respectively a and b.[9] Setting technical details aside, we can talk about the total spin of a set of particles, and the different values of each particle's spin depend of the direction we are measuring it. The two experimenters A and B each use a magnet that can be set to measure the component of the spin of the particle arriving toward them along one or another direction. In particular,

now is. First I must see how it looks when it is purified of instrumentalist frivolity and dares to say something not just about pointer readings but about the constitution of the world; and when it is purified of supernatural tales about the power of observant minds to make things up" (Lewis 1986, p. xi).

[9] One should not take this language too seriously, but for what is relevant to this discussion, one can imagine a particle like a spinning magnet, and think of its spin as its magnetization, so that we can measure the spin of the particle using a suitable magnetic field.

experimenter *A* can perform an experiment on *a* to determine its spin along three orthogonal directions *(x,y,z)*; and experimenter *B* can perform an experiment on *b* to determine its spin along the direction *w*.

1.2.1 The Axioms

Quantum theory predicts that the possible results for these experiments are constrained so that only certain values can come out from the measurements. This is "the SPIN axiom: Measurements of the squared (components of) spin of a spin 1 particle in three orthogonal directions always give the answers 1, 0, 1 in some order" (Conway and Kochen 2006, p. 227). That is, the set of results obtained by *A* on *a* is always one of the triplets 1,1,0; 1,0,1; or 0,1,1; and the result obtained by *B* on *b* is always either 0 or 1.[10]

In addition, quantum mechanics predicts that it is possible to produce pairs of 'twinned' particles, whose individual spin properties are interconnected with one another. In other words, they are *entangled* particles. This is "the TWIN axiom: For twinned spin 1 particles, suppose experimenter *A* performs a triple experiment of measuring the squared spin component of particle *a* in three orthogonal directions *x, y, z*, while experimenter *B* measures the twinned particle *b* in one direction, w. Then if *w* happens to be in the same direction as one of *x, y, z*, experimenter *B*'s measurement will necessarily yield the same answer as the corresponding measurement by *A*" (Conway and Kochen 2009, p. 228). That is, whenever $w = x$ (respectively *y*, or *z*) then the outcome obtained by *B* coincides with the first (respectively second, or third) digit of the result obtained by *A*. That is, *A* and *B*'s results are perfectly correlated. Like SPIN, also TWIN is a consequence of the formalism of quantum mechanics. To emphasize that, in the following I will write QM = SPIN & TWIN.

The third assumption of the theorem does not come from quantum mechanics but from relativity theory. Since it deals with the finiteness of the velocity of light, the axiom is called "FIN" (from the first three letters of 'finite'): "there is a finite upper bound to the speed with which information can be effectively transmitted" (Conway and Kochen 2009, p. 1443). Angelo Bassi and GianCarlo Ghirardi (2007) as well as Roderich Tumulka (2007) have argued that FIN is equivalent to a locality condition, namely the assumption that events in a region of space do not affect events in a region which is space-like separated from it. If so, these authors claim, Conway and Kochen's result is another instance of Bell's theorem (Bell 1964) which shows that no local theory can correctly reproduce the predictions

[10] Actually, SPIN is not properly an axiom but rather a theorem (Kochen and Specker 1967), so that if quantum mechanics is correct, the results of such spin measurements have to be constrained as SPIN says.

of quantum mechanics.[11] Therefore, these authors claim that this Conway and Kochen's theorem, based on the false FIN assumption, is unsound. In response, Conway and Kochen (2009) reformulated the Free Will Theorem. They dub it the 'Strong Free Will Theorem' and they use another axiom instead of FIN, called MIN: "assume that the experiments performed by *A* and *B* are space-like separated.[12] Then experimenter *B* can freely choose any one of the [...] directions *w*, and *a*'s response is independent of this choice. Similarly and independently, *A* can freely choose any one of the [...] triples *x, y, z*, and *b*'s response is independent of that choice" (Conway and Kochen 2009, p. 228). That is, when performing an experiment on the two twinned particles moving in opposite directions, there is always a minimum time the information needs for traveling from one particle to the other (hence the name "MIN"). In other words, MIN says that the experimental outcomes for *a* are independent of what experiment *B* chooses to perform on *b* and vice versa. To simplify the notation since both FIN and MIN are a consequence of relativity, in the following, I will denote either FIN or MIN with R.

In addition to SPIN, TWIN and MIN, there is another assumption in the theorem, namely that the outcomes of the experiment performed on one of the particles functionally depend on the previous state of affairs. That is, there are two function, *Fa* for particle *a* and *Fb* for particle *b,* each of which expresses the results in terms of the initial state. This functional dependence is Conway and Kochen's definition of determinism (thus the assumption's name "DET"): "particle *a*'s response is a function [...] of the information [...] available to it" (Conway and Kochen 2006, p. 1445).

1.2.2 The Proof of the Strong Free Will Theorem

Here is my reconstruction of the proof. Assume DET: there are functions, *Fa* and *Fb*, connecting the experimental outcomes with the initial states and which express the results of the experiments on *a* and *b* respectively. Because of MIN, each of these results, say F*a*, does not depend on the experiment that *B* actually performs on *b*. Given SPIN and TWIN, *Fa* and *Fb* can assume only a certain range of values, and a particular relation between the two particular functions, called '101-functions, holds.[13] Conway and Kochen provide a geometrical proof that such 101-functions in the current experimental setting cannot exist (Conway and Kochen 2006, p. 1468).

[11]Even if Tumulka, Ghirardi and Bassi believe that FIN is exactly the locality condition required in Bell's proof, there is a vast literature that discusses the various notions of locality: see (Redhead 1989) for a review. Moreover, there is no full agreement on what Bell's theorem proves, as also remarked in footnote 15.

[12]That is, the space distance between the two events is too large for a light signal emitted at one event to reach the other event, so that one event cannot cause the other. [This footnote is present in the original text.]

[13]The details of these functions are irrelevant for our purposes.

Therefore, it is impossible to have outcomes of experiments to functionally depend on previous states of affairs in ways consistent with the axioms QM (=SPIN, TWIN) and R(=MIN):

$$(1)\ \ (\text{QM} \ \&\ \text{R})\ \&\ \text{DET} \rightarrow \text{contradiction.}$$

To solve such contradiction one should reject one of the premises. Conway and Kochen argue that SPIN and TWIN, being at the heart of quantum mechanics, should not be rejected. Similarly, MIN being a consequence of relativity theory, is also difficult to deny. Therefore, they argue, the only option is to reject DET. Denying DET, Conway and Kochen continue, amounts to say that particles are free. That is, using an obvious notation in defining FW_particles = ~ DET, we have the "Strong Free Will Theorem" for deterministic theories (strong Free Will Theorem for deterministic theories, or sFWTd):

$$(\text{sFWTd}) \qquad (\text{QM} \,\&\, \text{R}) \rightarrow \text{FW\_particles.}$$

In their words: "the axioms SPIN, TWIN and MIN imply that the response of a spin 1 particle to a triple experiment is free—that is to say, is not a function of properties of that part of the universe that is earlier than this response with respect to any given inertial frame" (Conway and Kochen 2009, p. 228).

Conway and Kochen additionally claim that "randomness won't help" (Conway and Kochen 2006, p. 1463). In fact, they propose a method of converting any stochastic model into a deterministic one: "let the stochastic element [. . .] be a sequence of random numbers (not all of which need be used by both particles). Although these might only be generated as needed, it will plainly make no difference to let them be given in advance. But then the behavior of the particles in such a theory would in fact be a function of the information available to them (including this stochastic element) [. . .]" (Conway and Kochen 2006, p. 1463). This analogy from Tarun Menon (2010) is helpful to understand the proposal: suppose you and your friend want to play a game of die, then you roll all dice before the game, write down all the results, and then use this fixed information to play the game. Conway and Kochen claim that we would still have the same sort of functional dependence (denoted with DET' in the following equation) that gives rise to the contradiction noted above (assuming TWIN, SPIN and MIN are preserved in the stochastic-deterministic conversion):

$$(2)\ \ (\text{QM} \ \&\ \text{R})\ \&\ \text{DET'} \rightarrow \text{contradiction.}$$

Therefore, the more general version of the strong Free Will Theorem, valid also for stochastic theories reads as follows:

$$(\text{sFWTd} \,\&\, \text{i}) \qquad (\text{QM} \,\&\, \text{R}) \rightarrow \text{FW\_particles.}$$

That is, it is a consequence of quantum mechanics and relativity that particles are free. If the theorem really proves this, it is very good news for the libertarian: not only nobody could say that their view is contrary to physics, but also they would have a mathematical proof that they are correct! Unfortunately, however, I will argue in Sect. 1.4 that this is too good to be true. Nevertheless, before this, let me discuss in the next section the other criticisms that the theorem has received in the literature.

1.3 Criticisms 1: The Constraints on the Interpretations of Quantum Mechanics

In addition to this, Conway and Kochen take their theorem to be an "impossibility proof" against deterministic completion of quantum mechanics and the possibility of constructing relativistic invariant stochastic quantum theories. Quantum mechanics suffers from the measurement problem: if quantum mechanics is a theory only about the wave-function evolving according to the Schrödinger evolution, then unphysical "macroscopic superpositions," that is in superpositions of macroscopically different states of affair (like a dead and an alive cat), arise. Several theories have been proposed to deal with this, some of which are deterministic, like the pilot-wave theory (de Broglie 1928; Bohm 1952), others instead are stochastic, like the spontaneous localization theory (Ghirardi et al. 1986). The former avoids macroscopic superpositions postulating that the complete description of any physical system is given by the wave-function together with the particles position evolving deterministically. The spontaneous localization theory instead postulates that the wave-function evolves stochastically so that the macroscopic superpositions promptly disappear.

Conway and Kochen argue that their theorem rules these theories out. Since they claim that conditional (1) implies that determinism is false, they conclude that deterministic completions of quantum mechanics are impossible. Moreover, from conditional (2) they conclude that any stochastic completions of quantum mechanics so constructed cannot be made relativistic invariant, given that they would violate relativity (by violating MIN).

1.3.1 MIN in Deterministic Theories

Several authors have criticized these claims. Goldstein et al. (2011a) argue that MIN, like FIN, is equivalent to a locality condition, LOC, which entails that the (probability) distribution of the experimental results for a is independent of the distribution of the results for b. LOC, according to these authors, can be broken down in two conditions: parameter independence (PI) and outcome independence (OI). PI says that the experimental outcomes for a are independent on the parameters chosen by B for the experiment to perform on b; OI says that the results for a are independent on the results for b. For deterministic theories, since there is just one outcome, OI is trivially true and LOC reduces to PI. Since, according to these authors, in this case PI = MIN, then LOC = MIN. Therefore (sFTWd) reads:

$$(\mathrm{QM}\ \&\ \mathrm{LOC})\ \&\ \mathrm{DET} \rightarrow \text{contradiction.}$$

The problem is that, according to these critics, Bell's theorem shows that:

$$\text{(BT)} \qquad (\text{QM} \,\&\, \text{LOC}) \rightarrow \text{contradiction},$$

so that the contradiction in (sFTWd) is not resolved by rejecting DET, as Conway and Kochen do: one would have to reject either LOC or QM. Thus, the critics conclude, the sFWTd should be considered as an additional proof on nonlocality on par of Bell's theorem. As such, then, it does not pose any particular threat to the pilot-wave theory or any other deterministic completion of quantum mechanics.[14]

Christian Wüthrich (2011) similarly claims that if DET is true, then outcomes of one arm of the experiment will depend on the settings of the other arm. Therefore, for any deterministic theory that violates PI, and therefore MIN, (sFWTd) does not apply.[15]

1.3.2 MIN in Stochastic Theories

Considering now (sFWTd&i), the theorem works only if, as Conway and Kochen propose, there is a method to convert any stochastic theory into a deterministic one, "putting all randomness into the past" (Goldstein et al. 2011a, p. 1455). However, Goldstein et al. (2011a) show that such method would make MIN false, and therefore would invalidate the conclusion that DET' is false. In fact, assuming MIN reduces to PI, then PI is violated by the conversion method proposed by Conway and Kochen because "if nature were to follow the recipe suggested [...] then she should have to use the values of $k=k(x,y,z,w)$ depending on both experimenters' choices, (x,y,z) and w, in order to produce any of the outcomes" (Goldstein et al. 2011a, p. 1455). Therefore, according to these authors, the contradiction in (Conway and Kochen 2009) is again resolved because MIN is false, and not because DET' is false.

Goldstein and collaborators note that Kochen has suggested that MIN should not be interpreted as PI but rather "as requiring that the actual outcome itself of [one experiment] to be independent of B's choice, and not just its probability distribution." However, Tumulka (2016) writes that this strategy of pre-generating random

[14]That deterministic quantum theories like the pilot-wave theory must violate parameter independence has been known for a long time, but apparently the fact has not been appreciated enough.

[15]Notice that critics disagree on what Bell's theorem proves: while (Bassi and Ghirardi 2007; Tumulka 2016; Goldstein et al. 2011a) as well as (Albert 1992; Maudlin 1994) claim that it proves nonlocality, i.e. ~LOC, (Menon 2010; Wüthrich 2011) instead seems to think that it rules out local deterministic completions of quantum mechanics, i.e. ~(LOC&DET). If it is the former, then Bell's theorem provides a constraint for all quantum theories: any quantum theory (deterministic or stochastic) has to deny locality. In contrast, if it is the latter, Bell's theorem provides constraints only to deterministic quantum theories, and not on stochastic ones. Luckily, this distinction is not relevant from the discussion in this paper. For a discussion of the relation of Bell's theorem and the Free Will theorem, see Cator and Landsman 2014).

information will not work for his proposal of a relativistic invariant spontaneous collapse theory. This is because the distribution of the flashes, the ontology of the theory, depends on the choice of the directions of both arms of the experiment. If the distribution were given in advance, also these choices must be given in advance and they cannot because of DET'. In response, Conway and Kochen (2009) change their argument. Instead of pre-generating the information about flashes (which depends on the particular choice), they pre-generate the flash distribution of all possibilities. In this way, the choice together with the pre-generated information determines the particles' response. However, Menon (2010) claims that that would be a violation of (sFWTd&i) since the particles' response would not be independent of past information and thus they would violate DET'. Menon therefore identifies the real problem to be MIN. He believes that it incorporates a notion of "robust" causation, which is too strong. Similarly, Wüthrich (2011) argues that Conway and Kochen's argument against Tumulka either is an illegitimate way of introducing randomness, or it is legitimate but then it defeats itself.

1.4 Criticisms 2: Which Constraints the Free Will Theorem Actually Poses of the Free Will Debate?

Regardless of whether one considers the criticisms presented in Sect. 1.3 to be decisive or not, there are other concerns regarding the theorem's impact on the philosophy of free will. Therefore, let us assume for the sake of the argument that sFWTd&i works. If so, then physics (quantum mechanics and relativity) entails that there is free will in a libertarian sense, since the proof involves the denial of determinism. As emphasized, if true, that would be great news for the struggling libertarians. Let us see whether this is the case.

1.4.1 Begging the Question

Going back to the definition of MIN, one immediately sees that there is an additional assumption we haven't spelled out: "[...] experimenter *B* can *freely* choose [...]"(Conway and Kochen 2009, p. 228, emphasis added). Thus, the core of (sFWTd&i) is that *if* the experimenters have free will, *then* also the particles are free. That is, the conditional strong Free Will theorem is:

$$\text{(Cond.sFWT)} \quad \left(\text{QM} \,\&\, \text{MIN}'\right) \&\, \text{FW\_people} \rightarrow \text{FW\_particles},$$

where FW_people is the assumption that experimenters have free will, and MIN' is the portion of MIN without such assumption. However, if so, the theorem begs the question: the problem for the philosopher interested in free will is to determine whether the experimenter has free will! Formulated in this way, therefore, the theorem loses much of its appeal to the libertarian philosopher.[16]

1.4.2 The Conditional Claim

Proving the conditional claim might nevertheless be interesting. However, if the criticisms in Sect. 1.3 are correct, Conway and Kochen do not manage to prove it. In fact, (Cond.sFWT) implies that:

$$(\mathrm{QM}\ \&\ \mathrm{MIN})\ \&\ \mathrm{DET} \rightarrow \text{contradiction.}$$

For deterministic theories, if R = MIN = LOC, we have

$$\mathrm{QM}\ \&\ \mathrm{LOC}\ \&\ \mathrm{DET} \rightarrow \text{contradiction,}$$

that together with Bell's theorem (BT) implies that LOC is false, not that DET is. Hence, one cannot conclude that FW_particles = ~ DET is true. In addition, deterministic models of indeterministic theories fail MIN, so that

$$(\mathrm{QM}\ \&\ \mathrm{MIN})\ \&\ \mathrm{DET}' \rightarrow \text{contradiction}$$

implies that MIN is false, not that DET is. So again, we cannot conclude that FW_particles is true.

1.4.3 On the 'FW_people' and the 'FW_particles' Conditions

Even granting for the sake of the argument that the criticisms reported in Sect. 1.3 are mistaken, there are further problems connected to the fact that the notion of freedom for people discussed by Conway and Kochen is far from clear. If the result is (Cond.sFWT), is the assumption FW_people true?

Conway and Kochen say that FW_people is *presumably* true because it is the denial of determinism, and determinism is an implausible view, just like solipsism: "both the non-existence of free agents in determinism and the external world in solipsism are rightly conjectured up by philosophers as consistent if unbelievable universes to show the limits of what is possible, but we discard them as serious views of our universe" (Conway and Kochen 2006, p. 1462).

[16] Also Wüthrich (2011) claims that the theorem is question begging, even if in a different way: while Wüthrich is concerned on whether the Conway and Kochen theorem proves indeterminism, I am more concerned in whether it proves free will, and the literature on free will teaches us that the relation between lack of determinism and free will is not straightforward.

However, consider what FW_people says: "experimenters are free to choose between possible experiments" (Conway and Kochen 2009, p. 228). This sense of freedom is not incompatible with a deterministic universe: even if there is just one possible future, the experimenter does not know which one it is. Therefore, for all relevant purposes, one just needs an *epistemic* rather than a *metaphysical* notion of freedom. FW_people could therefore assert that the world is *as if* the experimenter can choose of orienting the magnet along a given direction. Since this is compatible with a deterministic universe, FW_people is not necessarily the denial of DET: it could just be a compatibilist notion of freedom. That is, even a compatibilist version of the FW_people assumption would do the trick for Conway and Kochen. Conway and Kochen, though, do not consider this possibility, since they regard determinism as "not serious:" they therefore want a libertarian notion of freedom. The problem though, is that they provide no argument for it: they simply write that if determinism is true then there is no way of making sense of science. However, this is not the case. Their worry seems to be that, if determinism is true then it would be pointless for an experimenter to perform experiments. If so, though, they are conflating predictability in practice with predictability in principle: if determinism is true it is possible in principle to predict the results of all possible experiments, but that does not mean that the experimenter actually has (or has to have) the necessary information to perform such computation. Thus, compatibilist freedom seems to be enough to make sense of science. Indeed, Landsman (2017) has similarly argued that the notion of freedom in the Free Will Theorem is a compatibilist one.

In addition, the theorem is advertised as showing that "if indeed there exist any experimenters with a modicum of free will, then elementary particles must have their own share of this commodity" (Conway and Kochen 2006, p. 1444). In other words, "if experimenters have a certain property then spin 1 particles have *exactly the same* property. Since this property for experimenters is an instance of what we usually call 'free will,' we find it appropriate to use the same term also for particles" (Conway and Kochen 2006, p. 1444, emphasis added). That is, FW_people is the same property as FW_particles.

However, we have just seen that FW_people is not necessarily the denial of DET (since it can be compatible with it), while FW_particles is, by definition. Thus, even if FW_people is true, the theorem does not show that the same property applies to people and particles.

Nevertheless, for the sake of the argument, assume that "FW" means the same thing for particles and people. Now the question is whether it is possible for particles and people to share a property like free will. Even if philosophers like Alfred Whitehead (1929) have arguably entertained such a view, it seems an implausible one: how can the property "FW" really mean what we ordinarily mean by free will and at the same time be attributed to people and to particles? This seems to involve a category mistake: while it seems appropriate (at least, intuitively) to consider observers as agents which may possess properties like free will, beliefs, desires, or knowledge, it does not seem to be sensible to ascribe these properties to particles, which are not agents, and whose typical properties are position, momentum, mass, or spin.

1.4.4 The Free State Theorem: Doing Without the 'FW_people' Condition

Let us set these considerations aside for a moment and consider the role the FW_people assumption plays in the proof of the Free Will Theorem. I argued above that it might simply express the idea that the world is as if different experiments can be performed, not that the world has actually an open future, which is perfectly compatible with a deterministic universe with a compatibilist free will. Because of this, it seems to me, the FW_people assumption should not be necessary in the proof. It is beyond the scope of this paper to see whether this is truly the case.[17] Interestingly enough, Conway and Kochen seem to recognize that: "there is a modification of the theorem that does not need the Free Will assumption. Physical theories since Descartes have described the evolution of a state from an initial arbitrary of 'free' state according to laws that are themselves independent of space and time. We call such theories with arbitrary initial conditions *free state theories*" (Conway and Kochen 2009, p. 1447). Consequently, they propose a version of the Free Will Theorem that does not contain the FW_people assumption, which they call "The Free State Theorem," FST (Conway and Kochen 2006, p. 1447):

$$\text{(FST)} \qquad (\text{QM}\,\&\,\text{MIN}) \rightarrow \text{FW\_particles}.$$

Conway and Kochen observe that it would be extremely unpleasant if the theorem would depend on FW_people. In fact, as many before them,[18] they speculate that "it is natural to suppose that this latter freedom [of the particles] is the ultimate explanation of our own" (Conway and Kochen 2009, p. 230). However, if the Conditional String Free Will theorem (Cond.sFWT) theorem works at best it proves the opposite, namely that people's freedom grounds the freedom of the particles. Therefore, they need to prove FW_particles independently of FW_people, which they allegedly do with the FST.

Therefore, if the FW_people assumption is not needed and the FST is sound, we might have arrived to something interesting, namely that particles are free, regardless of whether we are. Nonetheless, does the FST prove that particles are free? Landsmann (2017) has argued that the theorem should be seen from a compatibilist perspective. However, as already noted, Conway and Kochen explicitly reject compatibilism (since they reject determinism), so that they have to go with a libertarian notion of free will for particles. However, we have already seen that this is difficult to define: the core idea of libertarians is to attribute free will to people as agents and never to particles, in virtue of the fact that agents have properties that particles do not possess. In other words, in the libertarian framework

[17] See (Menon 2010; Wüthrich 2011; Norsen 2017; Bell 1985; Clauser et al. 1985; Goldstein et al. 2011b; Maudlin 2014; Bricmont 2016; Tumulka 2007) and references therein for a relevant discussion in the context of Bell's theorem.

[18] See e.g. (Kane 1996).

we are fundamentally different from particles: in particular, particles have no free will, only we, as agents, do.

Another option for Conway and Kochen is not to invoke agency and to stick with their definition of FW_particles as the denial of determinism but still different from randomness. Indeed, Conway and Kochen suggest that 'free particles' means 'particles that are randomly behaving constrained by the axioms of quantum mechanics.' They write: "the freedom we have deduced for particles is more constrained, since it is restricted by the TWIN axiom" (Conway and Kochen 2009, p. 230). In other words, particles' behavior is not described functionally and it is constrained by TWIN, while randomness is behavior is completely without constraints. This is compatible with their assumption that FW_particles = ~ DET. However, I think that this does not help at all: constrained randomness is, for all our purposes, still randomness, and the traditional objections that randomness is not freedom still hold. This constrained randomness implies that the behavior of particles is governed by laws which put constraints on their random behavior. This means that particles are string puppets whose strings jerk randomly but, say, cannot exceed a certain limit. As in the case of pure randomness, particles are not in control of their behavior. If one wishes, one can call this 'freedom' but this notion has not much to do with our common understanding of freedom.

To conclude, I think that the impact of the theorem on the free will debate is disappointing. In fact, even if the theorem proves something about the property "FW_particles," such property is either inconsistent, given that libertarian freedom attributed to particles seems an oxymoron, or it is 'constrained' randomness, and as such it is difficult to see how it may have something to do with the concept of freedom as traditionally intended.

References

Albert, D. Z. (1992). *Quantum mechanics and experience*. Cambridge: Harvard University Press.

Balaguer, M. (2004). A coherent, naturalistic, and plausible formulation of libertarian free will. *Noûs, 38*(3), 379.

Bassi, A., & Ghirardi, G. C. (2007). The Conway–Kochen argument and relativistic GRW models. *Foundations of Physics, 37*(2), 169.

Bell, J. S. (1964). On the Einstein–Podolsky–Rosen paradox. *Physics, 1*, 195.

Bell, J. S. (1985). Free variables and local causality. *Dialectica, 39*, 103.

Bohm, D. (1952). A suggested interpretation of the quantum theory in terms of 'hidden' variables, I and II. *Physics Review, 85*, 166.

Bricmont, J. (2016). What did Bell prove? In M. Bell & S. Gao (Eds.), *Quantum nonlocality and reality: 50 years of Bell's theorem* (p. 49). Cambridge: Cambridge University Press.

Cator, E., & Landsman, K. (2014). Constraints on determinism: Bell versus Conway–Kochen. *Foundations of Physics, 44*, 781.

Clarke, R. (2003). *Libertarian accounts of free will*. New York: Oxford University Press.

Clauser, J., Horne, M., & Shimony, A. (1985). An exchange on local beables. *Dialectica, 39*, 97.

Compton, A. (1935). *The freedom of man*. New Haven: Yale University Press.

Conway, J. H., & Kochen, S. (2006). The free will theorem. *Foundations of Physics, 36*, 1441.

Conway, J. H., & Kochen, S. (2009). The strong free will theorem. *Notices of the American Mathematical Society, 56*, 226.
de Broglie, L. (1928). La nouvelle dynamique des quanta. In J. Bordet (Ed.), *Electrons et photons: Rapports et discussions du cinquième conseil de physique* (p. 105). Paris: Gauthier-Villars.
Earman, J. (1986). *A primer on determinism*. Dordrecht: Reidel.
Ghirardi, G. C., Rimini, A., & Weber, T. (1986). Unified dynamics for microscopic and macroscopic systems. *Physical Review D, 34*, 470.
Goldstein, S., Tausk, D. V., Tumulka, R., & Zanghì, N. (2011a). What does the free will theorem actually prove? *Notices of the American Mathematical Society, 57*(11), 1451.
Goldstein, S., Norsen, T., Tausk, D. V., & Zanghì, N. (2011b). Bell's theorem. *Scholarpedia, 6*(10), 8378.
Kane, R. (1996). *The significance of free will*. Oxford: Oxford University Press.
Kochen, S., & Specker, E. (1967). The problem of hidden variables in quantum mechanics. *Journal of Mathematics and Mechanics, 17*, 59.
Landsman, K. (2017). On the notion of free will in the Free Will Theorem. *Studies in History and Philosophy of Modern Physics, 57*, 98.
Lewis, D. (1986). *Philosophical papers II*. New York: Oxford University Press.
Loewer, B. (2003). Freedom from physics: Quantum mechanics and free will. *Philosophical Topics, 23*(2), 91.
Maudlin, T. (1994). *Quantum non-locality and relativity: Metaphysical intimations of modern physics*. Cambridge: Cambridge University Press.
Maudlin, T. (2014). What Bell did. *Journal of Physics A, 47*(42), 4010.
Menon, T. (2010). *The Conway-Kochen free will theorem, manuscript*.
Merali, Z. (2006). Free will – You only think you have it. *New Scientist and Science Journal, 190*(2550), 8.
Norsen, T. (2017). *Foundations of quantum mechanics: An exploration of the physical meaning of quantum theory*. Cham: Springer.
Norton, J. (2008). The dome: An unexpectedly simple failure of determinism. *Philosophy in Science, 75*(5), 786.
Nozick, R. (1981). *Philosophical explanations*. Cambridge: Harvard University Press.
O'Connor, T. (Ed.). (1995). *Agents, causes, and events: Essays on indeterminism and free will*. New York: Oxford University Press.
Penrose, R. (1994). *Shadows of the mind: A search for the missing science of consciousness*. Oxford: Oxford University Press.
Pinker, S. (1997). *How the mind works*. New York: Norton.
Popper, K. (1972). *Objective knowledge*. Oxford: Claredon Press.
Redhead, M. (1989). *Incompleteness, nonlocality, and realism: A prolegomenon to the philosophy of quantum mechanics*. Oxford: Clarendon Press.
Searle, J. (1984). *Mind, brains, and science*. Cambridge: Harvard University Press.
Stapp, H. (1991). Quantum propensities and the brain–mind connection. *Foundations of Physics, 21*(12), 1451.
Stapp, H. (1993). *Mind, matter and quantum mechanics*. New York: Springer.
Stapp, H. (1995). Why classical mechanics cannot naturally accommodate consciousness but quantum mechanics can? *Psyche, 2*(5).
Stapp, H. (2017). *Quantum theory and free will: How mental intentions translate into bodily actions*. New York: Springer.
Strawson, G. (1986). *Freedom and belief*. Oxford: Oxford University Press.
Tumulka, R. (2007). Comment on 'the free will theorem'. *Foundations of Physics, 37*, 186.
Tumulka, R. (2016). The assumptions of Bell's proof. In M. Bell & S. Gao (Eds.), *Quantum nonlocality and reality: 50 years of Bell's theorem* (p. 79). Cambridge: Cambridge University Press.
van Inwagen, P. (1983). *An essay on free will*. Oxford: Claredon Press.
Whitehead, A. N. (1929). *Process and reality*. New York: Macmillan.
Wüthrich, C. (2011). Can the world be shown to be indeterministic after all? In C. Beisbart & S. Hartmann (Eds.), *Probabilities in physics* (p. 365). Oxford: Oxford University Press.

Chapter 2
Mind and Matter

Marcus Appleby

Abstract In physics it is well-known that finding the right question is often at least 50% of the difficulty. This paper does not propose a solution to the problem of consciousness, not even a tentative one. Rather it examines the question. This is in the belief that the question is currently mis-posed. The current conception of consciousness may be regarded as an attenuated version of the Cartesian concept of mind. It is argued that the Cartesian philosophy was originally motivated by conceptual problems with Galilean physics. Quantum mechanics changes things. This is not to say that the problem of consciousness is a pseudo-problem, as is sometimes suggested. It is, however, to say that the problem is not quite as is often assumed. In particular, it is argued that current conceptions encourage an unbalanced conception of mentality, according to which the state of being barely awake is the essence of what it is to be human, whereas the thought processes which led Einstein to the general theory of relativity are something a zombie could manage.

2.1 Introduction

Broadly speaking people interested in the problem of consciousness fall into two groups, one group arguing that mind is completely reducible to ordinary physiological mechanism, the other that the mechanism must be supplemented with some additional ingredient. The two groups tend to speak past each other. As Hardcastle (1997) resignedly puts it, at the beginning of a paper on the subject,

> I . . . recognize that I have little convincing to say to those opposed to me. There are few useful conversations; there are even fewer converts.

M. Appleby (✉)
School of Physics, Centre for Engineered Quantum Systems, The University of Sydney, Sydney, NSW, Australia

Stellenbosch Institute for Advanced Study, Matieland, South Africa
e-mail: marcus.appleby@sydney.edu.au

J. Acacio de Barros, C. Montemayor (eds.), *Quanta and Mind*, Synthese Library 414, https://doi.org/10.1007/978-3-030-21908-6_2

As it happens Hardcastle is a member of the first group, but there are probably many on the other side who share her pessimism about the likelihood of changing any minds. This differs from the situation usual in science, where no matter how difficult the problem, and no matter how intense the controversy, there is an expectation that consensus will eventually be achieved. Some might argue that consensus is not to be expected in the case of consciousness, since it is a philosophical problem not a scientific one, and philosophy is notorious for never reaching consensus about anything. This paper is motivated by the more optimistic view, that the problem ought to be solvable to the satisfaction of every competent person, and that if it seems otherwise it is because we are not looking at it in the right way.

The situation is in some ways reminiscent of a trial conducted under an adversarial legal system, where one person is appointed council for the prosecution, another is appointed council for the defence, and then they have a fight. This may, perhaps, be a good way of arriving at the truth under conditions where the question is well-posed. But if the question is not well-posed—if the truth is not represented by either of the alternatives on offer—then the adversarial process will either result in an answer that is guaranteed to be wrong, or else it will result in stasis. It appears to me that "stasis" would be a fair characterization of the situation with the problem of consciousness. To get out of the stasis one needs to put on hold the attempt to find a solution, and instead to re-examine the question. Specifically, one needs to look for shared beliefs: assumptions which seem obvious, and which both sides consequently take for granted, but which are in fact mistaken. Einstein's formulation of the theory of relativity depended on his identifying just such an assumption: namely, the assumption that there is an absolute, observer-independent sense in which events are either simultaneous or non-simultaneous. Something similar is required here.

On the face of it a philosopher like Chalmers (1996) and a philosopher like Dennett (1991) are at opposite ends of a spectrum. I would suggest, however, that what unites them is at least as important as what separates them. The modern concept of consciousness, as it features in philosophical controversy,[1] is the intellectual descendant of the Cartesian soul. One group of thinkers, while rejecting dualism of the full-blooded seventeenth century variety, retain a belief in a kind of attenuated version of the Cartesian soul. Another group reject the concept in its entirety. This much is apparent, and it is what creates the impression of strong disagreement. What is rather less apparent is that both sides continue to be constrained by some of the assumptions which originally caused Descartes to propose his distinctive theory of mind. So long as those assumptions remain unaddressed no satisfactory resolution is possible.

I will argue that the problem of consciousness (as currently conceived) is intimately related to the quantum interpretation problem (as currently conceived). I do not mean by this that an appeal to quantum mechanics can make the peculiar features of consciousness (as currently conceived) seem any less peculiar.

[1] As opposed to, for example, uncontroversial medical textbooks.

Rather I mean that the mistaken assumptions which underlie our confusions about consciousness are closely connected to the mistaken assumptions which underlie our confusions about quantum mechanics. Except by specialists Descartes is nowadays chiefly remembered as a philosopher. However, in his own mind, and in the minds of his contemporaries, he was at least as much a physicist.[2] More than that: it was precisely Descartes' attempt to clarify and to systematize the foundations of classical physics which originally led him to his concept of mind.

In Sect. 2.2 I examine the seventeenth century origins of the modern concept of consciousness; in particular, its connections with seventeenth century physics. The assumptions then made seem subsequently to have acquired, in many people's minds, almost the status of self-evident truth. Viewing them in historical context is important, because it helps one to see that they are in fact highly questionable. In Sect. 2.3 I discuss the relevance of quantum mechanics. To someone indoctrinated with seventeenth century ideas quantum mechanics seems weird. But I would suggest that it is really classical physics which should be seen as weird, at least when classically interpreted, and that quantum mechanics marks, or at least should mark, the restoration of sanity. Finally, in Sect. 2.4 I discuss the implications for the problem of consciousness.

In the following I make no attempt to suggest a solution to the problem of consciousness. Nor do I suggest that it is a pseudo-problem. Quite the reverse, in fact. My aim is only to expose some hidden assumptions, which I believe may be blocking progress. Some of the ideas were developed at greater length in a previous paper (Appleby 2014), to which the reader is referred for more detail.

2.2 Historical Origin of the Modern Concept of Consciousness

Modern physics rests on a fusion between two apparently very different attitudes of mind: The empirical attitude, according to which nothing is to be accepted except what can be experimentally validated, and the Pythagorean attitude,[3] which conceives the world as essentially mathematical in character. Concerning the latter Galileo wrote

> Philosophy is written in this all-encompassing book that is constantly open before our eyes, that is the universe; but it cannot be understood unless one first learns to understand the language and knows the characters in which it is written. It is written in mathematical language, and its characters are triangles, circles, and other geometrical figures; without these it is humanly impossible to understand a word of it, and one wanders around pointlessly in a dark labyrinth. (*The Assayer* Galilei (2008), p. 183)

[2]Of course, in the seventeenth century physics was considered a part of philosophy. However, a distinction between physics and what seventeenth century thinkers called metaphysics was still recognized.

[3]See, for example, Burtt (1954).

This doctrine of Galileo may be seen as an early progenitor of mathematical realism. It provokes an obvious question. The world does not superficially *look* like a book written in the language of mathematics. How, then, does Galileo account for all its qualitative features? Following the ancient atomists (Furley 1987) he simply dismisses such features out of hand, as a subjective illusion:

> Accordingly, I say that as soon as I conceive of a corporeal substance or material, I feel indeed drawn by the necessity of also conceiving that it is bounded and has this or that shape; that it is large or small in relation to other things; that it is in this or that location and exists at this or that time; that it moves or stands still; that it touches or does not touch another body; and that it is one, a few, or many. Nor can I, by any stretch of the imagination, separate it from these conditions. However, my mind does not feel forced to regard it as necessarily accompanied by such conditions as the following: that it is white or red, bitter or sweet, noisy or quiet, and pleasantly or unpleasantly smelling; on the contrary, if we did not have the assistance of our senses, perhaps the intellect and the imagination by themselves would never conceive of them. Thus, from the point of view of the subject in which they seem to inhere, these tastes, odors, colors, etc., are nothing but empty names; rather they inhere only in the sensitive body, such that if one removes the animal, then all these qualities are taken away and annihilated. [*ibid*, p. 185]

I believe this proposition, that quantitative (or primary) properties like position or velocity are objectively real while qualitative (or secondary) properties like colour or taste "inhere only in the sensitive body" marks the origin of the modern concept of a *quale*, and along with it the modern problem of consciousness. Of course, Galileo was only reviving a doctrine first proposed 2000 years earlier. There is an important difference however: In the ancient world atomism was one speculative system among many, and anyone who did not like it could reject it with a clean intellectual conscience. But starting with Galileo the idea that qualitative properties are unreal has been invested with all the prestige of modern science. That being so it is worth noting the weakness of Galileo's original argument. The argument is clearly not empirical. In other places Galileo relies heavily on experiment and observation. But here he relies on reasoning alone (of course, it is hard to see how he could do otherwise: What kind of direct, experimental evidence possibly could show that an object which to all appearances *seems* red is not in fact *genuinely* red?). Moreover, the reasoning is hardly compelling. The fact that "my mind does not feel forced to regard" something as true is, by itself, no grounds at all for asserting that the something in question is false. Finally, Galileo makes no attempt to explain how processes in the "sensitive body" are supposed to produce the misleading impression of objectively existing qualities.

Although it is impossible to conclusively demonstrate, I believe there is reason to think that Cartesian dualism was originally motivated by Descartes' desire to develop the programme first adumbrated by Galileo and others: In particular, his desire to give a unified, mathematico-physical account of the universe as a whole, the "sensitive body" included. In other words, the motivation came from physics. It must be granted that this is not immediately apparent from what is probably Descartes' best known work, *Meditations from First Philosophy* (Descartes 1984). That work (from the second edition on) carried the subtitle "in which are demonstrated the existence of God and the distinction between the human soul and the body", and a dedicatory letter to the faculty of theology at the Sorbonne in which it is explained that the motive is the defence of religion. Moreover, the structure of the argument—the fact that it is laid out as a deductive argument, starting from the famous *cogito ergo sum*—suggests a demand for absolute certainty foreign to the much more modest epistemological demands of empirical science as we now understand it. For these and other reasons (Ryle 2009) was surely not alone in supposing that Cartesian dualism was motivated exclusively by Descartes' religious concerns, and was essentially unrelated to his scientific interests:

> Descartes found in himself two conflicting motives. As a man of scientific genius he could not but endorse the claims of mechanics, yet as a religious and moral man he could not accept, as Hobbes accepted, the discouraging rider to those claims, namely that human nature differs only in degree of complexity from clockwork. The mental could not be just a variety of the mechanical.

However, on closer examination one finds indications that the true state of affairs of affairs is exactly the opposite of this: that, so far from being peripheral, dualism is actually foundational to Descartes' conception of physics. Certainly that is how it is presented in his subsequent *Principles of Philosophy* (Descartes 1982). In a letter to Mersenne (Descartes 1991) dated 28th January 1641 he not only explicitly confirms the connection with physics; he also says that he has intentionally concealed it:

> ...and I may tell you, between ourselves, that these six Meditations contain all the foundations of my physics. But please do not tell people, for that might make it harder for supporters of Aristotle to approve them. I hope that readers will gradually get used to my principles, and recognize their truth, before they notice that they destroy the principles of Aristotle.

A third indication comes from an examination of the earlier part of his career. Posterity remembers Galileo for his contributions to physics while Descartes is remembered chiefly for his contributions to philosophy[4] and mathematics. However, that is not at all how it would have appeared at the time. Particularly during the ten years after 1618, when Descartes worked on a wide variety of individual physical problems in a manner that was often highly mathematical (Shea 1991; Gaukroger et al. 2000; Schuster 2013), a contemporary would have seen him and Galileo as colleagues engaged on the same enterprise. At the end of this period Descartes set about synthesising the individual insights he had gained into a single, comprehensive picture of the universe, and of ourselves within it, conceived along the lines of mathematical physics (as we now call it), the results of this synthesis being set out in *The World* (Descartes 2004). He did not complete this work, which was only published after his death. The reason is that in 1633 he learned of Galileo's condemnation by the Inquisition. His first thought was that he should "burn all my papers"; his eventual decision was not to burn them, but only to show them to select friends (letter to Mersenne, end of November 1633 [Descartes 1991]). This episode is important as it shows why Descartes might have wanted to conceal the fact that

[4]In the modern meaning of the word "philosophy" which differs, of course, from the seventeenth century meaning.

the *Meditations* are foundational to his physics, and why he sought to bring the Aristotelians (i.e. the defenders of official Church teaching) round slowly instead of confronting them directly.

I believe this early work is very suggestive of the actual genesis of his dualism.[5] In it there is no mention of the *cogito* argument, and no attempt to prove the existence of God. Instead Descartes sets out a unified, systematic account of the world conceived as a physical mechanism. In particular he gives an account of the human brain as a mechanism, not essentially different from those constituting the rest of the cosmos. He ends with a promise to give a description of the "rational soul". Unfortunately this part of the manuscript was either never written, or else has been lost. However, I believe it is clear from what goes before (for instance his statement on p.149 of Descartes (2004), about what the soul will experience when "united to this machine") that he intended to say that the soul is a separate, immaterial entity interacting with the brain *via* the pineal gland. The question is why he would have taken that view. As I noted in Appleby (2014) the Aristotelian position was that the soul is merely the form of the body. If Catholic theologians were content with that, it may at first sight seem puzzling that the physicist Descartes should have postulated a much more thoroughgoing separation of soul and body. However, I think one can see on further reflection that it was precisely Descartes' commitment to mechanical explanations which would have pushed him in that direction. In *The World* Descartes begins by arguing, like Galileo before him, that only primary properties like shape or position are objectively real. Consequently, when he came to describe the mind one of the things he had to explain was how the impressions of secondary qualities like colour or taste are subjectively generated. Here he faced a problem. His description of the internal workings of the brain is extremely detailed. It seems likely, therefore, that he would have felt the full force of the difficulty which Thomas Huxley describes in the following graphic terms

> . . . how it is that anything so remarkable as a state of consciousness comes about as a result of irritating nervous tissue, is just as unaccountable as the appearance of the Djin, when Aladdin rubbed his lamp in the story. (Huxley 1866, p.193)

If Huxley, and numerous others, have thought it impossible to explain how *qualia* could be generated mechanically, inside the brain, it seems reasonable to suppose that Descartes, who was perhaps the first to wrestle with the problem, would have thought the same. In a seventeenth century context, when the vast majority of

[5]I am not suggesting that it should be seen as a kind of fossil, preserving Descartes thinking complete and entire, exactly as it stood in 1633. On the one hand there is evidence (Machamer and McGuire 2009) that he continued to revise it for some years after 1633. On the other hand there is evidence (Hatfield 2014; Schuster 2013) that he had already constructed early versions of some of the arguments that appear in the *Meditations* in 1629. My proposal is only that his dualism was originally motivated, not primarily by religious or moral concerns, but rather by considerations internal to his physics. The relevance of *The World* is that in it he presents a dualistic account of human nature without any of the accompaniments which elsewhere may suggest a motivation coming from somewhere other than physics.

scientists were religious believers, the obvious solution would have been to locate *qualia* outside the physical universe, in an entirely different kind of substance.

It remains to say a few words about the argument as it is presented in the *Meditations*. In *The World* Descartes argues that only primary properties like length are objectively real. There is an obvious difficulty[6] with this: For if the senses are profoundly misleading with respect to tastes and colours, why should it be thought that they are any less misleading with respect to numbers and lengths? What is to keep us from complete solipsism? In *The World* Descartes avoids the problem by confining himself to the assertion that the world conceivably might be constructed in the manner he describes. In the *Meditations*, however, he confronts it directly. As we have seen Descartes, in his dedicatory letter, presented the *Meditations* as being primarily about "God and the soul". However, though he chose not to advertise the fact, what the book does incidentally achieve is to justify the mechanical conception of nature (provided, of course, one accepts the intermediate steps).

2.3 Quantum Mechanics

Galileo and Descartes were two of the thinkers chiefly responsible for the birth of mathematical physics. In the four centuries since the subject has developed enormously. Nevertheless one can recognize in their writings two principles which continue to be highly influential down to the present day: namely (1) the Pythagorean assumption, that objective reality is either identical to or at least completely describable by an abstract mathematical structure and (2) the reductionist assumption, that every physical phenomenon is completely explicable in micro-mechanical terms. From now on I will refer to these principles as the Galilean-Cartesian concept of physical reality. Moreover, although dualism as such has long since been abandoned by most scientists, a weakened version of it remains in the shape of the modern problem of consciousness. What the discussion in the last section reveals is the fact that the two are intimately connected. Physicists typically think of consciousness as a problem for psychologists and/or philosophers, and nothing to do with them. But in fact it is they who created the problem in the first place. For most mathematical physicists, now as then, qualitative properties like colour are a nuisance, which they want to get out of the way so as to be able to concentrate on the serious business of constructing mathematical theories. The twenty-first century concept of consciousness, like the seventeenth century soul, serves as a convenient receptacle in which they can be dumped and forgotten. It is important to ask, therefore, whether the Galilean-Cartesian conception is actually justified.

---

[6] Democritus was already aware of the difficulty, as shown by the passage in which he has the senses respond to his atomistic philosophy with the retort "Wretched mind, you get your evidence from us, and yet you overthrow us? The overthrow is a fall for you." (Taylor 1999, fragment D23).

Modern physiology puts it beyond question that secondary properties are in a sense subjective. For instance, there are conventionally said to be seven colours in the visible spectrum. Although the number seven is perhaps a little arbitrary, no one is able to distinguish, say, 10^6 qualitatively different colours. The reason for this has to do with properties of the human eye-brain system, not properties of the electromagnetic spectrum.

By contrast, the status of primary properties, and the Galilean-Cartesian conception generally, have always been highly problematic. Direct empirical support was virtually absent in the seventeenth century, and even by the late nineteenth century it remained controversial, as can be seen from Mach's embrace of neutral monism (Banks 2003), or the fact that in the 1890s Planck was sceptical about atoms, to the extent that Boltzmann could attribute to him the opinion that work on kinetic theory was a "waste of time and effort" (Kuhn 1978, pp.22–3; also see Krips 1986). It was only in the twentieth century that scepticism about atoms died away. Unfortunately, however, the same advances which finally vindicated micro-mechanical explanations also cast serious doubt on Galilean-Cartesian assumptions about what such explanations should be like. Indeed, one of the key papers (Einstein 1905) leading to the general acceptance of atomism was published by the same person, in the same year, as one of the key papers (Einstein 1905) in the development of quantum mechanics.

Quantum mechanics does not describe a particle's trajectory. Instead it merely provides a way to calculate probabilities for the particle to be observed at different positions. This should be compared with the traditional analysis of secondary properties like colour. On the Galilean-Cartesian conception a rose does not possess an objective quality of redness, but only a disposition to produce a sensation of redness in us when one looks at it. Something similar is true of positions in quantum mechanics: a quantum particle does not possess an objective position, but only a probabilistic disposition to appear in various places when one observes it. One could express this by saying that in quantum mechanics every quantity is secondary. Moreover, mechanical models lose their status, as a fundamental description of the way things actually are, and become at best a heuristic device. Quantum mechanics thus challenges the Galilean-Cartesian conception at the most fundamental level. Although it is now almost a century since its discovery there is still no agreement how we should respond to the challenge. At one end of the spectrum there are those who think that giving up on the Galilean-Cartesian conception of reality amounts to giving up on reality altogether. They have accordingly tried to construct interpretations such as those of Everett (Saunders et al. 2010) or Bohm (Bohm and Hiley 1993; Holland 1993) which preserve the classical vision of the universe laid out before us, in a single conspectus, as it might appear to the mind of God. At the other end of the spectrum there are those like the QBists (Fuchs et al. 2014; Fuchs and Schack 2015; Fuchs 2016, 2017; Fuchs and Stacey 2016; Mermin 2018), who reject the Galilean-Cartesian conception in its entirety. For myself, I do not pretend to know how quantum mechanics should be interpreted. However, although I do not agree with the QBists on every detail,[7] I share their conviction, that quantum mechanics is pushing us towards an entirely different, and ultimately much more satisfactory conception of physical reality.

To someone raised on classical assumptions, abandoning the idea of primary properties may seem like a cognitive disaster. That is not the case, however. No one would think that, because language is a human invention, therefore the things we say using it are not true. No more does the fact that visual perceptions generally, and perceived colour qualities in particular, are brain constructions detract from the truth of the information we acquire by using our eyes. Visual perceptions may be subjective. But they are no more subjective than, for example, the belief that the electric field vector at position $\mathbf{r}$ is $3\mathbf{i} - 4\mathbf{j} + 7\mathbf{k}$ Vm^{-1}—where by "belief" I mean actual brain state. As Einstein stressed (Einstein 1934, 1970), the theories of mathematical physics are "free inventions," in the same way that human language is a free invention. This does not detract from the ability of mathematical physics to make true statements about the world. But it does cast doubt on the idea that the constructions of mathematical physics convey a brand of truth which is in some way superior to that conveyed by the constructions of the visual system.

It is, of course, true that we can be misled by various kinds of visual illusion. But that is no more to the point than the fact that the measuring instruments in a physics laboratory sometimes malfunction. It is also true that I have updated Descartes' argument in various ways. However, I do not think his original, seventeenth century version of the argument is any more persuasive. Descartes's first complaint about secondary properties is that they do not "resemble" anything in the object. But then the quantitative statements of mathematical physics do not resemble anything in the object either—neither the ink marks on paper, nor the beliefs in our heads. Moreover, it is hard to see how it would help if there was a resemblance. If resemblance is good then presumably identity is better. But it is surely not the case that a person with rocks in their head understands rocks better than someone not so afflicted. Descartes' other complaint is that the perceptions of secondary properties are not "clear and distinct": By which he presumably means (for example) that there is no sharply defined point where the perception of redness ends and the perception of orangeness begins. However, one runs into essentially the same difficulty when reading an analogue meter. Digital meters are certainly convenient. But they hardly make physics more objective than it used to be, back in the days before modern electronics.

Perhaps what is really behind the idea, that the descriptions of mathematical physics are objective in a way that visual perceptions are not, is the belief the former kind of description is in a certain sense canonical. Before the development of modern science abstract, symbolic descriptions conveyed much less information than visual perceptions. Indeed, to this day it is a common saying, that a picture is worth a thousand words. It was therefore natural to take the visual perception

[7]For instance, according to QBism quantum mechanical calculations only concern one's *experience* of a white dwarf, whereas it seems to me they must be saying something about the star *itself*.

to be the canonical representation, against which verbal descriptions were judged. Effectively, reality was identified with the visual perception (supplemented with information obtained from the other senses). However, the development of mathematical physics gave us an symbolic mode of description which, unlike ordinary language, was superior to visual perceptions in terms of informational capacity. It therefore became natural to take the mathematical description to be the canonical description: in effect, to identify reality with the mathematical description.

Quantum mechanics does not change the fact that mathematical descriptions are capable of conveying more information than visual perceptions. However, in quantum mechanics information is never complete.[8] It is therefore no longer appropriate to think in terms of there being a single, canonical description. Galileo's book metaphor is profoundly misleading. There is no comprehensive mathematical description in the sky. The only descriptions around are the ones we humanly construct and which, being human, are necessarily partial.

Galileo and Descartes were completely right to caution us against the common-sense tendency, to identify our perception of an object with the object itself. But we need to go one step further, and resist the temptation (which they encouraged) to identify the scientific conceptualization with the object itself. Given a collection of snapshots, all of the same person, no one would think to ask which is the once-and-for-all *correct* or *canonical* picture. Similarly with scientific descriptions.

After 400 years the Galilean-Cartesian identification, of objects with their mathematical descriptions, has become so ingrained that denying the possibility of a complete description may seem tantamount to denying the existence of reality itself. This perception is one of the motivations for constructing ontological interpretations, such as Bohm's. The perception is false, however. No one, talking to a friend, doubts that there is much more to that person than they or anyone else could possibly say. But that does not mean they doubt the friend's existence.

Quantum phenomena are often described as "weird". It is to be observed, however, that if one goes into a quantum physics laboratory one never sees anything weird (one may, of course, see something surprising, or remarkable, but that was true of classical laboratories). The sense of weirdness only arises when one tries to interpret the phenomena in terms of one of the mechanical models favoured by seventeenth century physicists. What this really shows is, not that nature is weird, but that seventeenth century physicists were wrong: The universe is not a giant machine. Mechanical models might alternatively be described as mechanical analogies. For instance, classical kinetic theory compares a gas with a collection of billiard balls. The comparison works quite well, in many situations. However, as is usually the case with analogies, it does not work perfectly. The fact that it breaks down if pushed too far is the opposite of surprising. Why should one expect there to

[8]The information one actually possesses as a result of observation, as opposed to the information one may imagine oneself possessing when contemplating a hidden variables model.

be a perfect analogy between the behaviour of billiard balls, and the behaviour of a gas at the microscopic level?

2.4 Implications for Consciousness

Similarly with the sense that there is something weird about the basic facts of human experience: this too is an artefact of a set of mistaken seventeenth century assumptions. It is sometimes argued that quantum mechanics cannot be relevant to the problem of consciousness because the relevant brain processes are, to a high degree of approximation, calculable classically. That may or may not be true. However, it has nothing to do with the argument I am presenting here, which is that quantum mechanics undercuts the usual perception of paradox at a basic ontological level.

Leibniz (Leibniz and Strickland 2014) sought to refute the idea that a machine could “think, feel, and have perception” by imagining it increased in size so that one could walk around inside it. He observed that if one did so one “would see only parts pushing one another, and never anything which would explain a perception”. The intuition driving this and similar arguments continues to be one of the main motivations for the feeling that there is something paradoxical about the phenomenon of consciousness. In imagination one compares the cerebral wiring diagram with, say, the visual appearance of a flower-garden, and asks oneself “how could *this* possibly be contained in *that*?” Such arguments depend on the perception that the mechanical model is a complete and perfectly accurate picture of things as they really are. Take away the idea that the brain simply is a machine—downgrade the mechanical picture to the status of mere heuristic analogy—and the argument loses its force.

What physics supplies us with are incomplete mathematical descriptions. Suppose one has an incomplete description of Alice’s brain, coded-up as a bit-string. Suppose also that Alice’s verbal description of her subjective experiences in a flower garden has been coded-up as a bit-string. No one would argue that the second bit-string cannot contain a description of Alice’s experiences because it doesn’t look like a flower-garden. So why should we accept such an argument in respect of the first?

None of this is to say that there are not important problems to do with the phenomenon of consciousness. It is, however, to say that the problem is not to understand how a colourless mechanism can generate the subjective illusion of colour. It may be hoped that this will lead to a more balanced conception of human mentality, in which the focus is less on the perception of colour qualities, and more on such achievements as the discovery of relativity. However impressive the ability to discern colour qualities may be, most of us ask a little more of our friends than a high score on the Glasgow coma scale (Teasdale and Jennett 1974).

References

Appleby, D. M. (2014). Mind and matter: A critique of Cartesian thinking. In H. Atmanspacher & C. A. Fuchs (Eds.), *The Pauli-Jung conjecture and its impact today* (pp. 7–36). Imprint Academic. `arXiv:1305.7381`.

Banks, E. C. (2003). *Ernst Mach's world elements: A study in natural philosophy*. Dordrecht: Springer.

Bohm, D., & Hiley, B. (1993). *The undivided universe: An ontological interpretation of quantum theory*. London/New York: Routledge.

Burtt, E. A. (1954). *The metaphysical foundations of modern science* (2nd ed.). Mineola: Dover Publications.

Chalmers, D. J. (1996). *The conscious mind: In search of a fundamental theory*. New York: Oxford University Press.

Dennett, D. (1991). *Consciousness explained*. New York: Little, Brown and Co.

Descartes, R. (1982). *Principles of philosophy* (V. R. Miller & R. P. Miller, Trans. and Edited). Dordrecht: D. Reidel Publishing Company.

Descartes, R. (1984). *The philosophical writings of descartes* (Vol. 2) (J. Cottingham, R. Stoothoff, & D. Murdoch, Trans.). Cambridge: Cambridge University Press.

Descartes, R. (1991). *The philosophical writings of descartes* (Vol. 3). Cambridge: Cambridge University Press. Translated J. Cottingham, R. Stoothoff, D. Murdoch, & A. Kenny.

Descartes, R. (2004). *The world and other writings* (S. Gaukroger, Trans. and edited.). Cambridge: Cambridge University Press.

Einstein, A. (1905). Über die von der molekularkinetischen theorie der wärme geforderte bewegung von in ruhenden flüssigkeiten suspendierten teilchen. *Annalen der Physik, 322*, 549–560.

Einstein, A. (1905). Über einen die erzeugung und verwandlung des lichtes betreffenden heuristischen gesichtspunkt. *Annalen der Physik, 322*, 132–148.

Einstein, A. (1934). On the method of theoretical physics. *Philosophy of Science, 1*, 163–169.

Einstein, A. (1970). Reply to criticisms. In P. A. Schilpp (Ed.), *Albert Einstein: Philosopher-scientist* (The library of living philosophers, Vol. VII, 3rd ed.). La Salle: Open Court.

Fuchs, C. A. (2016) Participatory realism. `arXiv:1601.04360`.

Fuchs, C. A. (2017). Notwithstanding Bohr, the reasons for QBism. *Mind and Matter, 15*, 245–300.

Fuchs, C. A., & Schack, R. (2015). QBism and the Greeks: Why a quantum state does not represent an element of physical reality. *Physica Scripta, 90*, 015104.

Fuchs, C. A., & Stacey, B. C. (2016) QBist quantum mechanics: Quantum theory as a hero's handbook. `arXiv:1612.07308`.

Fuchs, C. A., Mermin, N. D., & Schack, R. (2014). An introduction to QBism with an application to the locality of quantum mechanics. *American Journal of Physics, 82*, 749–754.

Furley, D. (1987). *The Greek cosmologists* (Volume 1: The formation of the atomic theory and its earliest critics). Cambridge: Cambridge University Press.

Galilei, G. (2008). *The essential Galileo* (M. A. Finocchiaro, Edited and Trans.). Hackett Publishing Company, Inc.

Gaukroger, S., Schuster, J., & Sutton, J. (Eds.). (2000). *Descartes' natural philosophy*. New York: Routledge.

Hardcastle, V. G. (1997). The why of consciousness: A non-issue for materialists. In J. Shear (Ed.), *Explaining consciousness—the 'Hard Problem'*. Cambridge: The MIT Press.

Hatfield, G. (2014). *The Routledge guidebook to descartes' meditations*. London: Routledge.

Holland, P. R. (1993). *The quantum theory of motion*. Cambridge: Cambridge University Press.

Huxley, T. (1866). *Lessons on elementary physiology*. New York: Macmillan and Company.

Krips, H. (1986). Atomism, Poincaré and Planck. *Studies in History and Philosophy of Science, 17*, 43–63.

Kuhn, T. S. (1978). *Black-body theory and the quantum discontinuity: 1884–1912*. New York: Oxford University Press.

Leibniz, G. W., & Strickland, L. (2014). *Leibniz's monadology: A new translation and guide.* Edinburgh: Edinburgh University Press.

Machamer, P., & McGuire, J. E. (2009). *Descartes's changing mind.* Princeton/Oxford: Princeton University Press.

Mermin, N. D. (2018). Making better sense of quantum mechanics. `arXiv:1809.01639`.

Ryle, G. (2009). *The concept of mind* (sixtieth anniversary ed.). London/New York: Routledge. With a critical discussion by J. Tanney.

Saunders, S., Barrett, J., Kent, A., & Wallace, D. (Eds.). (2010). *Many worlds? Everett, quantum theory, and reality.* Oxford: Oxford University Press.

Schuster, J. (2013). *Descartes-Agonistes: Physico-mathematics, method and corpuscular-mechanism 1618–33.* Dordrecht: Springer.

Shea, W. R. (1991). *The magic of numbers and motion: The scientific career of René Descartes.* New York: Science History Publications.

Taylor, C. C. W. (1999). *The atomists Leucippus and Democritus: Fragments, a text and translation with a commentary.* Toronto: University of Toronto Press.

Teasdale, G., & Jennett, B. (1974). Assessment of coma and impaired consciousness: A practical scale. *Lancet, 304*, 81–83.

Chapter 3
Between Physics and Metaphysics: A Discussion of the Status of Mind in Quantum Mechanics

Raoni Wohnrath Arroyo and Jonas R. Becker Arenhart

Abstract We discuss the 'Consciousness Causes Collapse Hypothesis' (CCCH), the interpretation of quantum mechanics according to which consciousness solves the measurement problem. At first, it seems that the very hypothesis that consciousness causally acts over matter counts as a *reductio* of CCCH. However, CCCH won't go so easily. In this paper we attempt to bring new light to the discussion. We distinguish the ontology of the interpretation (the positing of a causally efficacious consciousness as part of the furniture of reality) from metaphysics (the metaphysical character of that consciousness). That distinction allows us to map the philosophical theories of consciousness compatible with quantum mechanics under the tenets of CCCH. Also, it indicates that the problem will have to be discussed at a metaphysical level rather than at the physical level. Our analysis corroborates recent arguments to the effect that this interpretation is not ruled out so easily.

Keywords Consciousness · Interpretation of quantum mechanics · Ontology · Metaphysics

3.1 Introduction: Why Consciousness?

This volume deals with the relation between mind (or consciousness) and quantum mechanics (hereafter "QM"). Although some quite unrespectable claims are constantly made on behalf of such relation, it has undeniable pedigree: its sources are found on the earliest attempts to make sense of QM, when von Neumann (1955) put forth an interpretation based on the concept of a causal consciousness—frequently labeled as "Consciousness causes collapse hypothesis"

Supported by CAPES.

R. W. Arroyo · J. R. B. Arenhart (✉)
Research Group in Logic and Foundations of Science (CNPq), Graduate Program in Philosophy, Federal University of Santa Catarina, Florianópolis, Brazil

J. Acacio de Barros, C. Montemayor (eds.), *Quanta and Mind*, Synthese Library 414, https://doi.org/10.1007/978-3-030-21908-6_3

(hereafter “CCCH”). Remarkably, as recently shown by de Barros and Oas (2017), up to this date the CCCH has survived empirical tests as much as any other interpretation of QM; moreover, its specific features (namely, the introduction of a causal consciousness) cannot be independently subjected to an empirical falsification,[1] thus surviving as a live option to interpreters of QM.

However, as shown by the poll presented by Schlosshauer et al. (2013), the CCCH is rather unpopular among the theorists working on the field of the foundations of QM. As an illustration, in a recent book, Lewis (2016, §9) does not even consider the possibility of CCCH as a candidate for an interpretation of QM that could offer a reasonable worldview based on its commitments with dualism. It seems that having the label ‘dualism’ attached to it is enough for one to reject the CCCH. But what are the grounds for this rejection?

By rejecting CCCH because of its ties to dualism, what exactly are we rejecting? And for which reasons? The answer is not clear, it seems to us. The debates about mind and consciousness have always been problematic per se, and it should not be a surprise that when it comes to QM this issue only gets more complicated. In particular, difficulties arise from the fact that ‘dualism’ applies to a wide range of metaphysical options throughout the history of philosophy. As a result, even if it is clear that Neumann’s CCCH falls within the metaphysics of dualism, it is still far from clear *which* form of dualism the CCCH is committed with (e.g., the CCCH is dualist about substances or properties?).

In this paper we address precisely the issue of the nature of consciousness that is involved in CCCH. Given that CCCH is immune to empirical testing, this metaphysical approach is useful for both friends and foes of CCCH, and we take it as a first step towards a more rigorous investigation into the metaphysical basis of CCCH. We begin by arguing that, unlike many other interpretations of QM, the CCCH *determines* in large measure its ontology and, to a lesser degree, the metaphysical profile of its posits. That happens precisely because the CCCH is *incompatible* with several metaphysical profiles available in the philosophical literature on consciousness; following French’s Viking approach (French 2014) to the metaphysics of science, we argue that if one considers the distinct approaches to dualism available in the philosophical literature, most of them just won’t fit with CCCH. That puts the focus on some very specific approaches to dualism, making the theory and its ambitions clearer to evaluate. As a second step, given that dualism will be addressed metaphysically, we argue that there seems to be no good metaphysical reasons to rule out dualism *just because* it is dualist. To justify that claim, we will look at the metametaphysical literature to see whether dualism could be objectively ruled out in the face of other available metaphysical theories.

We structure the paper as follows. In the first section, we present the measurement problem and von Neumann’s (1955) solution to it. In the second section, we present a distinction between ontology and metaphysics that will enable us to somehow

[1]Nevertheless, it is not *unfalsifiable*, as the CCCH is *in principle* empirically distinguishable from any no-collapse approach to QM; see Ćirković (2005) for details.

extract an ontological commitment and determine a metaphysical profile to it. Besides, the ontology that the theory gives us is incompatible with many versions of dualism. In the third section, we will look at the literature on metametaphysics to see how dualism survives some of the typical arguments addressed against it. In the fourth section, we stress that CCCH is compatible with some rather specific kinds of dualism, in such a way that not every approach to consciousness in QM qualifies as a CCCH approach—thus reducing considerably the scope of the discussion between mind, causality, and QM. We conclude in section five.

3.2 The "Consciousness Causes Collapse Hypothesis"

The CCCH, as many other interpretations of QM, is essentially a *response* to the measurement problem. As there is much discussion about this problem in the literature, we will employ Maudlin's (1995) taxonomy, because it is a very concise way to put it. Moreover, we need only Maudlin's (1995, §1) "problem of outcomes" to see what is at stake here. The measurement problem, then, can be seen as the inconsistency of three basic assumptions of the wave-function representation of quantum states $|\psi\rangle$:

1.A $|\psi\rangle$ is *complete*: it specifies all the physical properties of the system it represents;
1.B $|\psi\rangle$ evolves *linearly* though time: its dynamics is described by linear equations of motion (e.g. the Schrödinger equation);
1.C Measurements of $|\psi\rangle$ always have determinate outcomes (e.g., either it is in one state or another, never in a superposition of states).

The proof of inconsistency of these three basic assumptions can be found in Maudlin (1995, pp. 7–8), so we will only comment it briefly. Suppose an experimentalist wants to measure the position of a quantum system S by means of a measurement apparatus A. According to 1.A and 1.B, the composite system $S + A$ evolves according to the Schrödinger equation. Then, by linearity, the states of $\hat{A}$ are also in superposition. This means that the possible A-states which correspond to, for example, different pointer positions, are superposed (hence, no definite single-state outcomes). But according to 1.C (and to our phenomenal perceptions in the laboratory and everyday life), we *do have* definite outcomes as a result of measurements. So, at least one of these assumptions must be dropped.

As we are interested in the CCCH approach, we are assuming implicitly von Neumann's (1955) formulation of QM, within the so-called "collapse approaches" to QM—which denies assumption *1.B*. The idea of "measurement" as the outcome of the interaction between a quantum signal (S) and a macroscopic measurement device (A) is, nevertheless, an intuitive one. So, in order to preserve this intuitive reasoning, one may suggest the attachment of a second measurement apparatus $\hat{A}'$ to measure the composite system $\hat{S} + \hat{A}$, in order to complete a measurement. We can reasonably consider it to be the experimentalist's eye, that observes the pointer reading. But as the second apparatus is related to the first apparatus in the same way that the first apparatus is related to the quantum object, linearity tells us that this is a new composite system $\hat{S} + \hat{A} + \hat{A}'$.

The problem remains in the sense that such interaction does not show a way out of the superposition describing the states of the composite system. In fact, this is the first step of an infinite regress, because one could suggest even that a *third* measurement apparatus is attached to the second apparatus, such as the optical nerve; and this optical nerve is related to a further measuring apparatus such as the brain, and so the argument goes *ad infinitum*.

This problematic situation is known as "von Neumann's chain". The main issue in von Neumann's measurement theory that we acknowledge here is that the linear descriptions of physical systems lead to an infinite regress: any attempt to reduce the superposition of the joint system with the introduction of further *physical* measuring apparatuses is doomed, since, as they are physical systems, they are to be described as a superposition as well. If the system is described by unitary dynamic laws, it will *always* be described by a superposition. As Baggott (1992, p. 186) stressed, it is difficult to fault the logic behind CCCH's conclusion: if the measuring device is a physical system, then it should be described by the equations of motion of QM as well as quantum systems are; moreover, if macroscopic physical measuring devices are composed by quantum systems, then they should, at least in principle, behave similarly; therefore, the superposition of macroscopic measuring devices' states (e.g., different pointer positions) is conceivable, and the interaction with the consciousness of the observer puts an end to the superposition's chain.

To von Neumann (1955, pp. 418–420), the solution of this problem is to recognize that the "*act* of measurement" takes place in the (subjective) perception of the observer, because one's subjective perception is the most reliable source that superpositions are not experienced at all. This feature of von Neumann's (1955, pp. 418–419) interpretation is known as the "principle of psychophysical parallelism" (see also Barrett 1999, §2.6).

To explain it, von Neumann (1955, p. 421) breaks down the measurement process into three stages: I, II, and III, where "I" is the quantum object, the system S being measured; "II" is the measurement apparatus (which could correspond to anything, from the instrument to the image registered in the observer's brain); "III" is *the observer*—more precisely, it is the observer's *abstract ego*.[2] The result of a measurement on I performed by $II + III$ is the same as the measurement on $I + II$ made by III. In the first case, the Schrödinger equation applies to I, and in the second case, it applies to $I + II$. That is, in all cases, the linearity of the Schrödinger equation does not apply to III, i.e., III is the only part in which a

[2]Although the term "consciousness" is absent, it is almost unanimous that von Neumann (1955, pp. 418–420) refers to the *consciousness* of the observer when he enunciates the causal feature of the "subjective perception" of the observer. For a historical motivation of this, see Jammer (1974, p. 480).

measurement occurs: it is only with the interaction of the abstract ego that the chain of superpositions collapses.

This agent is described as something outside the ontological domain of physical systems. As Becker (2004, p. 129) argues, the most relevant feature of the reasoning described above is that "physical processes must be explainable entirely in physical terms, but collapse, which is essential to the dual processes of quantum mechanics, cannot be explained entirely in physical terms". So the transition between a linear and a non-linear evolution is to be understood as a causal act of the observer's consciousness *upon* the composite system.

3.3 Ontology and Metaphysics

Here, we understand "ontology" as the study of *what there is*. Following the Quinean tradition, we assume that we can study scientific theories in order to extract the ontological commitments from those theories and discover *what there is* in the furniture of the world *modulo* such theories. That provides for a sort of *catalogue* of the beings the theory assumes as existing. That approach applied to CCCH allows us to claim that *consciousness* exists. Whatever it is in metaphysical terms, this entity—consciousness—is causally efficacious in the quantum measuring process. Consciousness is introduced in the furniture of the world by CCCH, with certain features (such as causal power).

We believe it is pacific that on what concerns CCCH's ontology consciousness exists. In the words of Ruetsche (2015, §3.3), "[w]hat a realist believes when she believes a theory T is an *interpretation* of T, an account of what the worlds possible according to T are like"; so the interpretation would provide for the realist content of the theory. In the case of QM, as it is well known, pragmatists do not generally think QM needs an interpretation. A theorist inclined to accept the CCCH must embrace consciousness and the role ascribed to it by the interpretation. But this is just as far as the theory leads us, in philosophical terms. That is, the interpretation forces the positing of consciousness, but the theory alone gives us no means to understand *what is* such consciousness in metaphysical terms. That is a pressing issue. Is it a fundamental property of all beings? Is it an emergent property? Is it a separated property? QM is silent about it. One simply cannot extract a metaphysical profile from an ontological catalogue accounting for what there is. So, if we want to inquire about that, we enter the domain of metaphysics. Here, we label this effort to come up with a "metaphysical profile" of the entities posited in a theory's ontology.

The idea that a metaphysical profile is needed in order to completely specify the realistic content of a theory was called by French (2014, p.48) *Chakravartty's challenge*. According to the challenge, it is not enough to point to some feature of a theory ('consciousness with causal powers' for instance) and say 'I am realist about that'. In order to have a legitimate realism about consciousness, one must clearly specify what it is, and doing so involves—at least partially—providing for a metaphysical characterization of consciousness. This links the ontology with the metaphysical profile. Also, providing for such profile may be enlightening, as we claimed in the introduction, if we are to know what the CCCH amounts to and to better ground any kind of attitude towards it we may happen to have (accept, reject, or whatever else).

As it happens, the metaphysical profile of the posits of scientific theories may be 'dressed metaphysically' in many incompatible ways, giving rise to a kind of metaphysical underdetermination. CCCH is much less liberal with the metaphysical profiles that may join the theory, as we shall see. In this sense, the ontology of a conscience with causal power requires a mind with very specific features. Let us see.

3.3.1 Dualism and Metaphysics

Traditionally, CCCH's consciousness is understood within a *substance-dualist* metaphysical profile (see Albert 1992, p. 83; Stapp 2011, p. 167; Stöltzner 2001, pp. 58–59). As we already mentioned above, in von Neumann's own solution to the measurement problem, the agent that causes the collapse is placed *beyond* the domain of application of QM, which concerns only the *physical* (in the sense of *material*) domain of reality. So, in this traditional viewpoint, consciousness acts upon the material domain causing the superposition of states to collapse into a non-superposed state; we find the quantum system in a definite state by virtue of such causal act of the observer's consciousness. This is a metaphysical statement about the nature and the behavior of this entity that was introduced in CCCH's catalogue of existing beings, and, in metaphysical terms, this is clearly a dualist claim.

But we should recognize that labeling CCCH "dualist" does not mean much if we are searching for the nature of the consciousness that is posited. There are many forms of dualism, and the CCCH is not compatible with every one of them. In this section, we will determine as much as possible the dualism(s) that may fit CCCH's ontology (so that one may address Chakravartty's challenge). On what follows, we will sketch some of the dualist taxonomy presented by Rodrigues (2014, pp. 201–203). This makes for a clearer case on what one is getting into by adhering to CCCH.

As one of its weakest formulations (Robinson 2017), we will define the basic dualism as *property dualism*, the thesis holding that the *material* properties (e.g., mass, charge, spin, and so on) and *mental* properties (e.g., consciousness, intentionality, qualia, and so on) are not reducible in terms of each other; moreover, the material and the mental are fundamentally of different nature. This is dualism at its most basic characterization, and other types of dualism can be distinguished by how they modify this basic thesis.

1. *Substance dualism*: Material and mental properties are different substances, and the bearer of such substances are also of a different nature;
    (a) *Strong substance dualism*: The mental stuff is immaterial, and its properties are distinct and exist independently of the material stuff;

(b) *Moderate substance dualism*: The mental stuff is immaterial, and its properties are distinct, but its existence depends on the material stuff;

As CCCH's ontology dictates, we will focus here on the *substance dualism*. Moreover, although the main difference lies in its strong and moderate versions, its taxonomy can be even further extended as:

1. *Pure dualism*: Material objects are defined by material properties only;
2. *Compound dualism*: Material objects are defined by material and mental properties;
3. *Non-spatial dualism*: Mental objects are defined by merely temporal properties;
4. *Spatial dualism*: Mental objects have spatial properties, hence, are extended through space;
5. *Theistic dualism*: Mental objects and properties are created by God;
6. *Naturalistic dualism*: Mental objects and properties are integrated with the material world;
7. *Interactionist dualism*: Material and mental objects maintain two-way causal relations;
8. *Epiphenomenalism*: Material stuff causes the mental stuff, but not the other way around;
9. *Pre-established harmony*: There are no causal relations between the material and the mental.

In this dualist spectrum, Cartesian dualism may be classified as a *strong theistic interactionist non-spatial pure dualism* (Rodrigues 2014, p. 203). But what about CCCH? Which one(s) of the above metaphysical taxonomy of dualism should apply to its ontology? As it seems, there are many consciousness-based approaches to the measurement problem in QM; we will briefly review what are those options and how well they fare (or don't) within the constraints imposed by CCCH.

In the interpretation presented by London and Bauer (1983), the consciousness of the observer is treated by another Hilbert subspace $\mathcal{H}_C$ that interacts with those of the "objective" parts. Moreover, as French (2002) has pointed out, in London and Bauer's (1983) theory of quantum measurement, consciousness is not *causal* in the quantum-mechanical process of measurement, but merely *recognizes* a measurement result, as a way of attributing meaning to it in a phenomenological way. Moreover, London and Bauer (1983) themselves recognize that their theory of measurement is to be understood within a phenomenological metaphysics (specifically a Husserlian approach), so it will not fit the consciousness *causes* collapse hypothesis. The *ontological* background is different from the one that we are investigating here. This issue deserves an investigation of its own, so we will not address the interpretation of London and Bauer (1983) here.

Another consciousness-based approach to QM is the *many-minds interpretation*, put forth mainly by Lockwood (1989), and it is easy to notice that it does not fit the causal aspect of consciousness of the CCCH: the main reason is that such approach is an extension of Everett's (1957) relative-state approach to QM, in which there is no collapse (thus, no agent whatsoever *causing* the collapse).

As it seems, it is just von Neumann's (1955) approach to QM that requires a consciousness with causal powers. So the basic ontological features of consciousness in CCCH are:

1. *Causality*: Consciousness must be a causal agent in the quantum measuring process;
2. *Transcendence*: The laws of QM that apply to *physical systems* should not apply to consciousness;
3. *Interaction*: There must be an interaction between physical systems and consciousness, as the latter modify the dynamics of the former.

These three main ontological features of CCCH's consciousness in fact show its incompatibility with several metaphysical profiles listed above. As it stands, the only metaphysical profiles that are compatible with CCCH's ontology are strong versions of naturalistic and interactionist dualism. All others are ruled out by some features of the ontology. Consider epiphenomenalism, for instance: it does not admit mental causation, so it is unable to count as an interpretation of CCCH's consciousness. It's incompatible with what the theory gives as an *ontological output*. The same would go to any moderate version of dualism as well: if the very existence of a substance, say, mental, is *dependent* of the material, then consciousness would not be able to act as a causal agent in the measuring process of QM; and the other way around would not be compatible as well, because the mind *alone* could not create a result of a quantum measurement—its causal power is strictly dependent of the experimental setup in which the quantum system lies in.

In this sense, we are now in position to look for some very specific attempts at determining what consciousness could be (and what it *couldn't*) according to CCCH. Obviously, that does not solve the problem, but it leaves us in a clearer situation than the one we began with (even to formulate more clearly the difficulties with the view). Notice also that once CCCH is adopted (even if only as a working hypothesis), this counts as our version of quantum mechanics. From this perspective, it is the theory that rules some metaphysical versions of dualism out, providing for a kind of epistemic authority that the metaphysics alone would not have (see also Arenhart 2012).

3.4 On Metametaphysics: Can We Rule Out Dualism on Metaphysical Grounds?

Even though we are able to classify with greater precision what CCCH's consciousness *could be* in metaphysical terms, dualism is still a very unpalatable idea for many. So perhaps it could be ruled out on other grounds, other than empirical? In this section we will address the debate from a metametaphysical perspective, searching for some kind of evaluation of metaphysical theories to see whether there are good arguments to rule out at least some of the above forms of dualism compatible with CCCH.

3.4.1 *Widen the Net*

If dualism could not be ruled out directly by physics (de Barros and Oas 2017), perhaps it could be ruled out if we expand our scope to other sciences which depend somehow on the results provided by physical theories, such as neurosciences. This metametaphysical criterion was recently coined by Benovsky (2016, pp. 82–84) as "widen the net": we should not look at isolated areas, but rather see how a metaphysical theory fits in a (more) general picture. In this sense, if dualism is compatible with physics, but incompatible with everything neuroscientists produced so far, the *widen the net* would be a good metametaphysical criterion to recommend that we abandon such metaphysical theory. However, this seems not the be case. As Arshavsky (2006) shows, there would not be a single result in neuroscience that would be incompatible with dualism (at least so far); in fact, his study shows that much of the neuroscientists' vocabulary is essentially dualist. So this metametaphysical criterion would not do when one is looking for a way to discard dualism.

3.4.2 *Causation*

If dualism is the only metaphysical profile that one is able to plug with CCCH's ontology, it could be argued that although von Neumann's proposal solves the measurement problem, it also raises other philosophically puzzling problems concerning mind-body causation. In fact, causation is the ground from which the traditional challenges to dualism often occur. So, if our best theories about causation are incompatible with dualism, then it would pose a problem to the interpreter of QM adhering to CCCH. However, as Rodrigues (2014, pp. 214–216) stressed out, the most popular theories about causation, such as counterfactual, covering law, probability raising, primitivist and energy flow theories of causation are all compatible with at least interactionistic versions of dualism—which the CCCH is also compatible with. So, still according to Rodrigues (2014, p. 84), "[w]hatever truth about causation is, the best theories we have now don't rule out immaterial minds causing bodily changes".

3.4.2.1 Causal Closure and Naturalism

Let us take into account another objection that is commonly held against CCCH, which is the violation of the causal closure of the world. Roughly, the causal closure thesis asserts that every physical event must have a physical cause, and if it is true, it is violated by the attribution of causal power to a non-physical entity (see Auletta and Wang 2014, p. 263). The argument can thus be written as:

1. Everything happens according to the laws of physics;
2. There is no mental causality in the laws of physics;

∴ There is no mental causality in the world.

Notice that the second premise is based on *naturalism*, the thesis which holds that science is our best guide to metaphysics. If it is right (and that is debatable), then one may add CCCH with mental causality to the laws of QM, hence denying this very step. So, the causal closure cannot be used to rule out a metaphysical thesis implied by a physical theory.

3.4.3 Uninformativeness

Another common ground of criticisms against dualism is uninformativeness (Rodrigues 2014, pp. 203–207): it is often objected that dualism does not adequately characterize, in metaphysical terms, what the mind-stuff *is*. As pointed out by de Barros (2014, §2), it seems that von Neumann's solution replaces "[...] a mystery by another mystery, without adding any explanatory power". It does not explain consciousness in terms of what it *is*, but in terms of what it *does*. In this sense, the CCCH is uninformative and hence should lose its attractiveness to interpreters of QM.

To resist this objection, one might look at *opposite* metaphysical views. Take materialism, for example. Does it answers what matter *is*? Its answer is as functional as the dualist's. In QM, other approaches to the measurement problem fail to explain what the mechanisms of measurement *are*: what are parallel universes (DeWitt 1970)? Or what is the mechanism responsible to physical collapses (Ghirardi et al. 1986)? The answer, again, is functional. CCCH is not worst off than the alternatives.

As pointed out by Rodrigues (2014, p. 222), dualism raises more questions than answers. But so does QM when relating to the measurement problem. So, there seems to be no definitive objections to CCCH and its dualist metaphysics.

3.5 Conclusion

Dualism suffers from a curious fate. While it seems a rather natural step in explaining conscious phenomena, and won't go away by any empirical means, it is widely regarded as way too exoteric in order to account for quantum collapse. In fact, there are rarely arguments against it; it is just taken by many to be a non-starter. To make matters worse, dualistic understandings of quantum mechanics are responsible for most of the pseudo-scientific literature on quantum mechanics, making it difficult sometimes to provide a sensible account of the view without prejudices.

We have attempted to provide a clearing of the ground for further serious work on the relation of mind and quantum mechanics on the specific interpretation first presented by von Neumann (the CCCH). Carefully articulating the view requires that the role of consciousness in QM is seen as a causal factor responsible for the collapse. In this sense, the ontology associated with CCCH requires a conscience with causal powers over matter, and that matter and mind be clearly distinguished. This, on its turn, provides for both a precisification on the kind of approach to consciousness that is compatible with CCCH as well as for a restriction on the scope of the metaphysical theories available to do the job. That illustrates a collaborative work between science and metaphysics, with science providing for a test ground for metaphysical theories.

Furthermore, we have argued that, given that metaphysics will play a major role in dressing the posited consciousness with some important features, purely metaphysical arguments are also—at least so far—unable to rule CCCH as implausible. Given that, it could be the case that CCCH could be much better understood if current forms of dualism compatible with it could be more clearly articulated together, so that existing metaphysical theories could be employed to somehow enlighten the role of consciousness in CCCH. This is a demanding task, we leave it for a future work.

References

Albert, D. Z. (1992). *Quantum mechanics and experience*. Cambridge: Harvard University Press.

Arshavsky, Y. I. (2006). "Scientific roots" of dualism in neuroscience. *Progress in Neurobiology, 79*(4), 190–204.

Arenhart, J. R. B. (2012). Ontological frameworks for scientific theories. *Foundations of Science, 17*(4), 339–356.

Auletta, G., & Wang, S.-Y. (2014). *Quantum mechanics for thinkers*. Singapore: CRC Press.

Baggott, J. (1992). The meaning of quantum theory: A guide for students of chemistry and physics. New York: Oxford University Press.

Barrett, J. A. (1999). *The quantum mechanics of minds and worlds*. Oxford: Oxford University Press.

Becker, L. (2004). That von Neumann did not believe in a physical collapse. *The British Journal for the Philosophy of Science, 55*, 121–135.

Benovsky, J. (2016). *Meta-metaphysics: on metaphysical equivalence, primitiveness, and theory choice* (vol. 374). Synthese Library. Cham: Springer.

Ćirković, M. M. (2005). Physics versus semantics: A puzzling case of the missing quantum theory. *Foundations of Physics, 35*(5), 817–838.

de Barros, J. A. (2014). On quantum mechanics and the mind. To appear in the *Proceedings of the Foundations of the Mind Conference*, Berkeley. http://userwww.sfsu.edu/barros/publications/publications/files/deBarros2014a.pdf

de Barros, J. A., & Oas, G. (2017). Can we falsify the consciousness-causes-collapse hypothesis in quantum mechanics? *Foundations of Physics, 47*(10), 1294–1308.

DeWitt, B. S. (1970). Quantum mechanics and reality. *Physics Today, 23*(9), 30–35.

Everett, H. (1957). "Relative state" formulation of quantum mechanics. *Reviews of Modern Physics, 29*(3), 454–462.

French, S. (2002). A phenomenological solution to the measurement problem? Husserl and the foundations of quantum mechanics. *Studies in History and Philosophy of Science Part B: Studies in History and Philosophy of Modern Physics, 33*(3), 467–491.
French, S. (2014). *The structure of the world: Metaphysics and representation*. Oxford: Oxford University Press.
Ghirardi, G. C., Rimini, A., & Weber, T. (1986). Unified dynamics for microscopic and macroscopic systems. *Physical Review D, 34*(2), 470.
Jammer, M. (1974). *The philosophy of quantum mechanics: The interpretations of quantum mechanics in historical perspective*. New York: Wiley.
Lewis, P. J. (2016). *Quantum ontology: A guide to the metaphysics of quantum mechanics*. New York: Oxford University Press.
Lockwood, M. (1989). *Mind, brain and the quantum: The compound 'I'*. Basil: Blackwell.
London, F., & Bauer, E. (1983). The theory of observation in quantum mechanics. In: J. Wheeler & W. Zurek (Eds.), *Quantum theory and measurement* (pp. 217–259) (J. Wheeler & W. Zurek, Trans.). Princeton: Princeton University Press.
Maudlin, T. (1995). Three measurement problems. *Topoi, 14*(1), 7–15.
Robinson, H. (2017). Dualism. In: E. N. Zalta (Ed.), *The Stanford encyclopedia of philosophy*. Fall 2017. Metaphysics Research Lab, Stanford: Stanford University.
Rodrigues, J. G. (2014). There are no good objections to substance dualism. *Philosophy, 89*(2), 199–222.
Ruetsche, L. (2015). The Shaky Game +25, or: On locavoracity. *Synthese, 192*(11), 3425–3442.
Schlosshauer, M., Kofler, J., & Zeilinger, A. (2013). A snapshot of foundational attitudes toward quantum mechanics. *Studies in History and Philosophy of Modern Physics, 44*, 222–230.
Stapp, H. P. (2011). *Mindful universe: Quantum mechanics and the participating observer*. Berlin/Heidelberg: Springer Science & Business Media.
Stöltzner, M. (2001). Opportunistic axiomatics—von Neumann on the methodology of mathematical physics. In: M. Rédei & M. Stöltzner (Eds.), *John von Neumann and the foundations of quantum physics* (pp. 35–62). Dordrecht: Springer.
von Neumann, J. (1955). *Mathematical foundations of quantum mechanics* (R. Beyer, Trans.). Princeton: Princeton University Press.

Chapter 4
Bridges Between Classical and Quantum

Leonardo P. G. De Assis

Abstract The use of quantum theory to model brain functioning has been the subject of much controversy, and some of the objections presented have discouraged many to invest in this area of research. In spite of these objections there is a great variety of basic physics researches, that although not directly applied in the modeling of the mechanisms of the brain, offer different possibilities of the use of quantum formalism to help understand the mechanisms behind the functioning of the brain. In this work, I will make a brief review of this literature and present evidence that there are many opportunities for the association of quantum theory and brain research.

Keywords Quantum theory · Brain physics · Quantum computing · Foundations of quantum mechanics

4.1 Introduction

Since the creation of quantum theory many hypotheses have suggested a relation between quantum phenomena and brain functions. The oldest one comes from the times of foundation of quantum mechanics, and was motivated by the need to explain the problem of measurement. Other motivation to suggest a parallel between brain processes and quantum theory was to explain how the brain is able to process information fast despite the neurons are known to be very slow compared to the processors used in digital computers. More recently another reason was motivated by research in cognition and decision making that showed that certain cognitive processes could only be properly modeled using the mathematical formalism of quantum theory. These findings raised the question whether the brain would also function according to quantum theory. Indeed, these quantum theories of mind have been the subject of recurrent debates (Tarlac and Pregnolato 2016), where different

L. P. G. De Assis (✉)
Stanford University, Stanford, CA, USA

J. Acacio de Barros, C. Montemayor (eds.), *Quanta and Mind*, Synthese Library 414, https://doi.org/10.1007/978-3-030-21908-6_4

arguments have been proposed to anchor this relationship between quantum physics and brain. Most of the time physical arguments were used to show the unfeasibility of those proposals. While these refutations focus on some aspects of these proposals for using quantum theory to model processes in the brain, they have had the side effect of driving away new research in that area.

In this paper I will briefly review the literature in physics to show that many of the objections to quantum theories of mind are not definite impediments, and that these various theoretical and experimental results found in the physical literature demonstrate the possibility of quantum phenomena in hot systems, as well as other experiments and theories that show the possibility of existence of quantum phenomena in macroscopic scales. I would argue that a careful consideration of the consequences of these finds may suggest other ways of use quantum theory to help unravel some of the mechanisms still unknown in the brain. Another aspect that I propose to highlight with these examples is that the use of quantum theory to explain brain functions is not only a matter of applying quantum concepts to neuroscience, but there are other ways to discuss the relationship between quantum theory and mental processes.

4.2 Classical and Quantum: Size

4.2.1 Quantum as Microscopic Phenomenon

It is common to find in introductory texts the definition of quantum theory as being a theory that deals with phenomena found on a microscopic scale. For example, among many sources with similar definition, we have the following definition of quantum mechanics according to the Stanford Encyclopedia of Philosophy (Ismael 2000):

> Quantum mechanics is, at least at first glance and at least in part, a mathematical machine for predicting the behaviors of microscopic particles... (Stanford Encyclopedia of Philosophy)

Although it is correct regarding what states, this definition is incomplete, and therefore gives room for misunderstandings. It is true that quantum theory describes microscopic phenomena, and it is also a historical fact that these theories were able to describe many phenomena that until recently were not observed at macroscopic scales. However, from these observations, we cannot extrapolate and say that quantum theory only describes microscopic phenomena. Within the similar habit of presenting quantum theory as describing microscopic phenomena, it is also common to say that the theory of relativity has as scope the physics of objects at very fast motion (near speed of light), or as the physics at the cosmological scale. This is also not true. In fact, only employing relativity, we can adequately describe most of subatomic processes. This way of introducing the main physical theories has led many people to mistakenly split the physical world into three different size scales. According to this widely held view, even among some scientists, quantum theory describes phenomena at microscopic scales, classical mechanics the events at our human size scale, and theory of relativity as responsible for cosmological scale physics.

Despite this widespread opinion, according to physics, quantum theory is a formalism that describes fundamental properties of the physical world, and therefore its consequences cannot be limited to a single scale of size. However, it is possible to understand the origin of this misconception by recognizing that each one of these theories is capable to better describe phenomena that we often observe only in each of these different size scales.

4.2.2 Quantum as Microscopic a Phenomenon

This distinction between the scales of application of the main physical theories has been very important objection in the debate about a possible relation between quantum phenomena and brain processes. Although since the foundation of quantum theory many physicists, including some founders of quantum theory, considered one possible relation between mind and quantum physics, recently this possibility gained new contours when some researchers proposed that the brain could work in a way similar to those theorized for quantum computers. One of these theories of the mind that uses quantum mechanics as the basis for explaining brain processes (Hameroff and Penrose 1996) argues that substructures that are part of neurons, called microtubules, are so small that quantum effects become unavoidable. To put into perspective a possible contribution of microtubules, it would be good to have the idea of the size of these structures compared to other objects. Since our central question is whether the brain could compute using quantum physics, it would be interesting to begin by comparing the microtubules with the transistors used in our computers. In this case we have that the transistors in our current computers have a size of approximately 14 nm. If we consider that the microtubule has approximately 200 nm – 25 μm (25,000 nm) length, this means that the microtubule will be 10^1 to 10^3 times larger than the transistors in use today. Other interesting comparisons would be with the size of HIV virus, and with the diameter of the silicon atom. In the first case, our transistors are 10 times smaller than the HIV virus, and at least 1000 times larger than the diameter of the silicon atom (0.2 nm). On the other hand, our current transistors are only 70 times larger than the diameter of the silicon atom. From this comparison we can see that the microtubules are no so small as a first impression might suggest. Another conclusion is that if quantum effects would be important in the size scale of microtubules, even more pronounced these effects would be for transistors. Although current processors do not perform quantum computation, transistors have their functioning based on quantum mechanics, but the risk that the quantum effects affect the behavior of these devices in the past was avoided due to the large size they had compared to today's transistors. However, since the size of today's transistor is on the scale of nanometers. Many operating failures due to quantum effects (mainly quantum tunneling) have been reported in the last ten years, and the industry has taken steps to minimize these effects. It is also important to remember that these quantum effects have been happening on computers, even considering that these devices dissipate a lot of energy in the form of heat. Of course, the existence of quantum effects doesn't imply the existence of appropriate conditions for some form of quantum computing happen. My intention here was to introduce the notion that the question of size for consider quantum effects has a relative importance. That small objects, with the size at which quantum effects would be important, can operate classically. In the following sections, I will show that large objects, on the human size scale, can exhibit quantum behavior.

4.2.3 Entanglement

Penrose's theory was received with skepticism by part of the academic community, which used as one of the main reasons for refuting that theory, the same argument usually used against the possibility of quantum computing: decoherence. According to this argument, the brain would be a very "wet and hot" environment, and therefore the phenomenon of quantum entanglement, considered fundamental for the characterization the difference between classical and quantum computation, would have a very short existence. Therefore, that short life would make quantum computation impossible would be caused by the influence of the environment, via decoherence. This refutation was followed by a debate in which the parties argued that important aspects present for each side argumentation were not taken into consideration. This controversy gave rise to a general opinion, especially among those who did not follow closely that debate, that since the brain is a very noisy environment, and therefore the phenomenon of decoherence would be an insurmountable obstacle to any proposal of application of the quantum theory to elucidate unsolved problems in neuroscience. This criticism to a quantum theory of mind seems to be dominant, even after years of experiments, and observations that show situations with long decoherence times at room temperature, or other quantum properties at large scale.

In fact, in a sequence of experiments Julsgaard and collaborators have demonstrated the possibility of entanglement in large systems with long decoherence time, and separated by large distances (Julsgaard et al. 2001; Sherson et al. 2005). In particular in a paper published in 2001 (Julsgaard et al. 2001) they showed that a caesium gas sample with about 10^{12} atoms submitted to the pulse of light was able to maintain entanglement by approximately 0.5 ms. In another work from 2005 (Sherson et al. 2005) they show that two systems, each composed of 10^{11} caesium atoms at room temperature, and separated by a distance of 35 cm or more, exhibited entanglement.

4.2.4 Superposition

Another quantum phenomenon that is often associated with small size scales is quantum superposition. Superposition is one of the most characteristic phenomena of quantum theory, as the well-known Schrodinger's cat thought experiment attests. It is common in the texts that introduce this thought experiment, be accompanied by a warning that informs that it is just an analogy to processes that only occur in the microscopic world. Despite this characterization of superposition, several experiments have been proposed in recent years to test the possibility of quantum superposition occurring at distances, and timescales of everyday life (Carlesso and Bassi 2018; Nimmrichter and Hornberger 2013; Leggett 2002; Marquardt et al. 2008; Lee and Jeong 2011). In fact, Kovachy et al. (2015) used light-pulse atom interferometry to show the existence of quantum interference in wave packets separated by a distance of 54 cm that lasted approximately 1 s. In another experiment Gerlich et al. showed experimental evidence of quantum interference of large organic molecules composed of up to 430 atoms (Gerlich et al. 2011). These experiments suggest that we should reconsider the belief that quantum superposition would not have long life in macroscopic systems, and possibly learning from new experiments of this kind will teach us how to prepare macroscopic systems to exhibit quantum superposition with long lives.

4.3 Classical and Quantum: Spin

4.3.1 Spin in Macroscopic Systems

In the works cited earlier on experimental observation of entanglement in macroscopic systems (Julsgaard et al. 2001; Sherson et al. 2005), the authors used a macroscopicic version for spin, collective spin, as the central element of those experiments. Therefore, those works not only showed experimental evidence for macroscopic version of entanglement, but also for spin.

In a similar way, a recent experimental work by Holanda et al. (2018) the authors show the existence of spin in magnons and that the light scattered by the phonons is circularly polarized, thus demonstrating that the phonos have an instrisic property equivalent to the spin. At this point it would be appropriate to remind the reader that similar to entanglement, spin is a central concept in any proposed quantum computation.

4.3.2 Classical Nonlinearity

Teklu et al. (2015) have addressed the relation between nonlinearity and nonclassicality in nonlinear classical systems (oscillators). They were able to demonstrate through the negativity of Wigner function, which is frequently used as a measure of nonclassicality, the existence of a relation between nonlinearity and quantumness. More specifically, they show that the increase of nonlinearity in classical systems was accompanied by an increase in the appearance of non-classicality, as measured by the Wigner function, in classical systems.

Spin is another concept that is considered to be non-classical. However, a few years ago, I and my colleagues showed that a Born-Infeld electrodynamics, which is a nonlinear generalization of Maxwell's electrodynamics, exhibits spin in its classical version (Vellozo et al. 2009) with a deviation of approximately 5%.

4.4 Classical and Quantum: Entanglement

4.4.1 Entanglement is Necessary Condition for Classicality

As it is well known, the quantum theory was developed after the establishment of classical physics. In that context, the appearance of non-locality and entanglement was a counter-intuitive surprise that generated a great deal of resistance to the new theory, and still today is a reason for objection by some researchers. Those unusual properties gave rise to assumption that entanglement has no relation to classical physics, since classical theory was not capable of predict that phenomenon. However, we could try to reverse the question and ask if it is possible to construct a non-classical theory that possesses entangled states, and that at the same time it does not have a classical limit. That was the question that Richens et al. (2017) addressed, and their answer was that such a situation was not theoretically possible. They considered all non-classical theories that can decohere to classical theory, and their conclusion was that the theories that met that requirement are theories with entangled states, or classical theories. According to their results the existence of entanglement is a necessary condition for the existence of classical systems. Their conclusion also implies that a non-classical theory, that has as limit the classical physics, has the properties of the quantum theory that we know. Even more important, this result is evidence of an additional link between classicality and quantumness.

4.4.2 Synchronization as Classical Signature of Entanglement

Recently Witthaut et al. (2017) showed a direct link between classical synchronization and quantum entanglement. Many researchers in the past compared the synchronization that appeared in classical and quantum systems. In these studies, each regime (classical or quantum) was investigated separately and then a comparison between them was established. However, a way to directly establish the link between classical and quantum synchronization when the system make the transitions from one regime to another was only recently suggested by Witthaut et al. From the quantum version of an isolated many-body system they were able to formulate, via mean-field, the classical version that had the form of Kuramoto equations. They observed that the coupling parameter values for which the mean field approximation synchronized corresponded to situations in which the corresponding quantum version exhibited entanglement.

4.5 Classical and Quantum: Decoherence

As we have seen in previous sections the main objection to quantum processes in the brain is based on the concept of decoherence. Quantum decoherence is a physical mechanism that would explain the transition from the quantum to the classical regime. What is not often discussed is that, similarly to what happens with many physical models, there is no single model and interpretation of decoherence. In fact different visions of decoherence have been adopted over time, and the one that is more widely used today has limitations that are not often addressed. In this section I will address this subject by adopting as reference two papers (Fortin and Lombardi 2017; Castagnino et al. 2008) that duly discuss this topic. Interested readers may seek to know more by reading these works and the bibliographic references therein.

In two interesting works, S. Fortin, O. Lombardi and colleagues classify the different views on decoherence in three approaches that not only depict the different concepts of decoherence but also the problems that led to the need for reformulation.

4.5.1 First Approach

In this first period, it was proposed that the classical macroscopic properties would emerge by separating the quantum system into states that would carry the information of the quantum properties and states that would carry the information of the classical properties. Due to a process of thermal relaxation the quantum degrees of freedom would converge to the thermal equilibrium leaving the degrees of classical freedom. Although this proposal discarded the use of terms of interference, and the time predicted to reach the thermal equilibrium did not correspond to the known decoherence time.

4.5.2 Second Approach

According to the approach suggested in the second period, decoherence is the result of the interaction between the system, and the environment. The basic hypothesis of this proposal is that the quantum systems are always in contact with the environment, and thus was able to predict a time of decoherence with a shorter time than the previous proposal. The environment-induced decoherence is still the "orthodox" view, and the one used to refute a possibility of quantum effects in the brain.

4.5.2.1 Second Approach: Difficulties

Despite being the orthodox version of decoherence, it presents some limitations that are not usually commented upon when this definition of decoherence is introduced. Among these limitations are:

1. Cannot be applied to closed systems, in particular to the universe as a whole;
2. Does not provide a criterion to decide where is the frontier between the quantum system and the environment;
3. Unable to define the classical behavior of the systems that emerge from the original quantum system.

4.5.3 Third Approach

Despite the success of decoherence induced by the environment and the fact of still the most widely spread proposal, other hypotheses have been proposed to explain situations where the inducedness by the environment hypothesis cannot explain, such as closed systems. One of these proposals suggests that decoherence in closed systems would be self-induced due to destructive interference.

4.6 Classical and Quantum: Formalism

4.6.1 Similar Formalism

In 2004 Prokhorov (2004) showed that it is possible to describe classical mechanics with the similar formalism used in quantum mechanics. According to that description the classical Hamiltonians have complex valued probability amplitudes. He also showed that in that reformulation there is one constant, phase-space area, with dimension of action, which would be the classical equivalent of Planck's constant, as well as a classical version of Fock's space. Although this reformulation is written using classical probability, it has the following additional hypothesis: determinism, physics at Planck distances, discreteness of the fundamental structure, and dissipative processes.

4.6.2 *Quantum Mechanics from Classical Statistical Physics*

In a series of papers (Wetterich 2009, 2010, 2018) Wetterich proposes that the dynamics of classical subsystems isolated from a larger system, which also includes the environment, reproduces the properties of quantum mechanics. Among the recovered properties are unitary evolution, noncommutativity, entanglement, quantum interference. He also proposes that such formalism can be implemented by neural networks, or by the brain (Wetterich 2018).

4.6.3 *Information Loss Approach*

Also using the dissipation hypothesis as a mechanism for quantization, Blasone et al. (2009) in 2009, generalized the 't Hooft quantization program for interacting systems. They showed that it was feasible to obtain a quantum isotonic oscillator from two classical Bateman's oscillators. In another parameter regime, the resulting quantum system could be interpreted as a particle in an effective magnetic field, interacting through a spin-orbit interaction term. If the dissipation of information was applied to each Bateman's oscillator separately it was possible to interpret the resulting quantum system as being two independent quantum harmonic oscillators.

4.7 Contextuality

According to De Barros and Suppes (2009) the contextuality would be the main quantum characteristic manifested in the brain, and that classical interference could explain why certain cognitive processes be better described by a mathematical formalism of quantum theory. In that work Barros points out that contextuality is a property present in both quantum and classical physics, and therefore we can avoid the hypothesis of the brain function according to the laws of quantum physics, to display properties associated with the mathematical formalism of quantum theory. In the same paper, Barros comments that the quantum entanglement is not a fundamental property when it comes to the brain, since given the distances, a signal can be propagated from one region to the other of the brain so fast that for practical purposes it can be considered as being instantaneous. This discussion about the importance of contextuality in the consideration of a quantum approach to mental processes draws attention to an important aspect not considered so far, which is about the benefits of using different interpretations of quantum mechanics. Implicitly this work has been assuming the most widely accepted interpretation of Copenhagen, yet since contextuality is a very important concept in the Broglie-Bohm weak interpretation, we might wonder if a different interpretation would help us to establish a mapping between processes in the brain and quantum processes.

4.8 Conclusion

In this work, I did not intend to present philosophical or neuroscience-based arguments to counter the main objections usually presented against quantum theories of the mind, I have just stated some results that can be found in the physics literature that contradict such objections. Despite these examples, I have been very far from exhausting all possibilities of using quantum theory to explain processes. For example, as I said at the beginning of this paper that one of the motivations was to explain the processing speed of information in the brain, yet another question that the quantum theory can shed some light on is how the brain can be energetically efficient. Quantum computing was the first example of an important current line of research called Reversible Computing. Here there is no suggestion that the brain is an adiabatic or quantum system, but that we can learn from the quantum reversible computation how to explain the energetic efficiency of the brain (Moret-Bonillo 2015). Even though this list is far from complete, I believe it constitutes a representative sample among many other works, which in a similar way show that there is at least reason to reconsider the objections that have traditionally been made against the use of quantum theory to model the physics of the brain.

Acknowledgements LPGA acknowledge support from the Patrick Suppes Gift Fund as well for the support from Byrne Gift Fund at Stanford University.

References

Blasone, M., Jizba, P., Scardigli, F., & Vitiello, G. (2009). Dissipation and quantization for composite systems. *Physics Letters A, 373*(45), 4106–4112.

Carlesso, M., & Bassi, A. (2018). Tests of the quantum superposition principle: Current experiments on Earth, future experiments in Space. In *Talk Presented at the Fifteenth Marcel Grossmann Meeting-MG15*, University of Rome, La Sapienza-Rome, 1–7 July 2018.

Castagnino, M., Fortin, S., Laura, R., & Lombardi, O. (2008). A general theoretical framework for decoherence in open and closed systems. *Classical and Quantum Gravity, 25*(15), 154002.

De Barros, J. A., & Suppes, P. (2009). Quantum mechanics, interference, and the brain. *Journal of Mathematical Psychology, 53*(5), 306–313.

Fortin, S., & Lombardi, O. (2017). A top-down view of the classical limit of quantum mechanics. In *Quantam structural studies: Classical emergence from the quantum level* (pp. 435–468).

Gerlich, S., Eibenberger, S., Tomandl, M., Nimmrichter, S., Hornberger, K., Fagan, P. J., Txen, J., Mayor, M., & Arndt, M. (2011). Quantum interference of large organic molecules. *Nature Communications, 2*, 263.

Hameroff, S. R., & Penrose, R. (1996). Conscious events as orchestrated space-time selections. *Journal of Consciousness Studies, 3*(1), 36–53.

Holanda, J., Maior, D. S., Azevedo, A., & Rezende, S. M. (2018). Detecting the phonon spin in magnon-phonon conversion experiments. *Nature Physics, 14*, 500–506.

Ismael, J. (2000). Quantum mechanics [Internet]. Stanford encyclopedia of philosophy. Stanford University [cited 24 Dec 2018]. Available from: https://plato.stanford.edu/entries/qm/

Julsgaard, B., Kozhekin, A., & Polzik, E. S. (2001). Experimental long-lived entanglement of two macroscopic objects. *Nature, 413*(6854), 400.

Kovachy, T., Asenbaum, P., Overstreet, C., Donnelly, C. A., Dickerson, S. M., Sugarbaker, A., Hogan, J. M., & Kasevich, M. A. (2015). Quantum superposition at the half-metre scale. *Nature, 528*(7583), 530.

Lee, C. W., & Jeong, H. (2011). Quantification of macroscopic quantum superpositions within phase space. *Physical Review Letters, 106*(22), 220401.

Leggett, A. J. (2002). Testing the limits of quantum mechanics: Motivation, state of play, prospects. *Journal of Physics: Condensed Matter, 14*(15), R415.

Marquardt, F., Abel, B., & von Delft, J. (2008). Measuring the size of a quantum superposition of many-body states. *Physical Review A, 78*(1), 012109.

Moret-Bonillo, V. (2015). Can artificial intelligence benefit from quantum computing? *Progress in Artificial Intelligence, 3*(2), 89–105.

Nimmrichter, S., & Hornberger, K. (2013). Macroscopicity of mechanical quantum superposition states. *Physical Review Letters, 110*(16), 160403.

Prokhorov, L. V. (2004). On physics at Planck distances: Quantum mechanics. *Physics of Atomic Nuclei, 67*(7), 1299–1311.

Richens, J. G., Selby, J. H., & Al-Safi, S. W. (2017). Entanglement is necessary for emergent classicality in all physical theories. *Physical Review Letters, 119*(8), 080503.

Sherson, J., Julsgaard, B., & Polzik, E. S. (2005). Distant entanglement of macroscopic gas samples. In V. M. Akulin, A. Sarfati, G. Kurizki, & S. Pellegrin (Eds.), *Decoherence, Entanglement and Information Protection in Complex Quantum Systems: Proceedings of the NATO ARW on Decoherence, Entanglement and Information Protection in Complex Quantum Systems*, Les Houches, 26–30 April 2004 (Vol. 189, pp. 353–372). Dordrecht: Springer Science & Business Media.

Tarlac, S., & Pregnolato, M. (2016). Quantum neurophysics: From non-living matter to quantum neurobiology and psychopathology. *International Journal of Psychophysiology, 103*, 161–173.

Teklu, B., Ferraro, A., Paternostro, M., & Paris, M. G. (2015). Nonlinearity and nonclassicality in a nanomechanical resonator. *EPJ Quantum Technology, 2*(1), 1–0.

Vellozo, S. O., Neto, J. H., Smith, A. W., & De Assis, L. P. (2009). Self-interacting electromagnetic fields and a classical discussion on the stability of the electric charge. *International Journal of Theoretical Physics, 48*(7), 1905–1911.

Wetterich C. (2009). Emergence of quantum mechanics from classical statistics. *Journal of Physics: Conference Series, 174*(1), 012008. IOP Publishing.

Wetterich, C. (2010). Quantum mechanics from classical statistics. *Annals of Physics, 325*(4), 852–898.

Wetterich, C. (2018). Quantum computing with classical bits. arXiv preprint arXiv:1806.05960. 15 June 2018.

Witthaut, D., Wimberger, S., Burioni, R., & Timme, M. (2017). Classical synchronization indicates persistent entanglement in isolated quantum systems. *Nature Communications, 8*, 14829.

Chapter 5
Where Does Quanta Meet Mind?

J. Acacio de Barros and Carlos Montemayor

Abstract The connection between quantum physics and the mind has been debated for almost a hundred years. There are several proposals as to how quantum effects might be relevant to understanding consciousness, including von Neumann's Consciousness Causes Collapse interpretation (CCC), Penrose's Orchestrated objective reduction (Orch OR), Atmanspacher quantum emergence theory, or Vitiello's field theory. In this paper, we examine the CCC, in particular Stapp's theory of interaction of mind and matter, and discuss how this imposes constraints to possible brain structures. We then argue that those constraints may allow us to identify a possible locus of the interaction between mind and matter, if CCC is true.

5.1 Introduction

In a seminal book, von Neumann formulated what became one of the main interpretational difficulties of quantum theory: the measurement problem (von Neumann 1983). In a nutshell, the measurement problem can be seen as a difficulty of understanding how the transition from the quantum to the classical domain happens, and, more importantly, how the quantum and classical dynamics and descriptions of nature can co-exist in a consistent way. Since von Neumann's book, there were many attempts to solve this problem, which led to diverse interpretations of the mathematical formalism, sometimes with very disturbing ontologies associated to them. Among the most well-known and discussed interpretations are the many worlds interpretation (Everett III 1957), Bohm's pilot-wave theory (Bohm 1952a,b), Nelson's stochastic theory (Nelson 1985), and epistemic interpretations (such as QBism; see, e.g., Fuchs 2014).

J. A. de Barros (✉)
School of Humanities and Liberal Studies, San Francisco State University, San Francisco, CA, USA
e-mail: barros@sfsu.edu

C. Montemayor
Department of Philosophy, San Francisco State University, San Francisco, CA, USA

J. Acacio de Barros, C. Montemayor (eds.), *Quanta and Mind*, Synthese Library 414, https://doi.org/10.1007/978-3-030-21908-6_5

There is no consensus among physicists or philosophers as to which of those interpretations is most adequate. However, there is perhaps no other interpretation that leads to a stronger negative reaction and even downright disdain by many physicists than the Consciousness Causes Collapse (CCC) interpretation of quantum mechanics, which invokes a different dynamics for quantum systems when they interact with a mind. This connection between quantum mechanics and the mind dates back to the beginning of the twentieth Century, when researchers were trying to understand the meaning of the quantum formalism. The idea was that, when looking at the difference in quantum and classical behavior of material particles, it was clear that we only observe classical behavior, but never quantum. For example, for a particle in a quantum superposition of two states, whenever observed (measured) for a property in a superposition, this particle would be seen as in either one or the other state, and never in a strange non-classical superposition of properties. This led von Neumann to postulate that the quantum realm had two different dynamics, one linear and deterministic, which would happen for isolated quantum systems, and another non-linear and probabilistic, which would happen when a quantum system interacted with the observer's mind (von Neumann 1983). The CCC interpretation seems to postulate a dualist view of the world: there is matter, which satisfies an unitary and deterministic evolution, such as Schroedinger's equation, including when interacting with material objects, such as a measuring device; and there is the mind, which not only is unclear as to which dynamics it follows, but also causes matter to evolve differently, via a non-deterministic and non-unitary quantum jump, when mind interacts with matter.

Whether we believe or not on the CCC interpretation, or whether we find its dualistic ontology too difficult to swallow, it is still an interesting question to ask what are its consequences, and whether we can learn something about quantum mechanics or the philosophy of mind by studying it. For example, Henry Stapp, a student of Emilio Segré who has devoted his life to the CCC interpretation, has developed a clever way to model the control of matter by a mind using the inverse Quantum Zeno Effect (IQZE). In this paper, we will explore this idea, and try to discuss what possible structures in the brain might be candidates for the loci of interaction between mind and matter.

This paper is organized as follows. In Sect. 5.2 we briefly review the measurement problem, and introduce some possible solutions, focusing more specifically in the CCC interpretation. In Sect. 5.3 we discuss Stapp's idea of using the IQZE to model the effects of the mind on the brain/matter. In Sect. 5.4 we discuss constraints to the behavior of matter, were the model presented in Sect. 5.3 true, and we argue that the search for the loci of interaction between mind and matter should look for certain particular aspects of brain dynamics. Section 5.5 ends this paper with some conclusions.

5.2 The CCC Interpretation

As we mentioned above, the CCC interpretation was the first attempt to address the measurement problem. So, it is fit to start our discussion of this interpretation with a discussion of this problem.

In quantum theory, the state of a physical system is described by a normalized vector in a Hilbert space. Measurements are represented by hermitian operators in the Hilbert space, called *observables*. Properties, the particular outcomes of an experiment, are then associated to eigenvalues of the observable, and states that are eigenvectors can be thought as having well-defined properties, since the repeated measurement of this property yields always the same outcome, i.e. the same eigenvalue and eigenvector.

Any vector on the Hilbert space can, in principle, be physically realizable through careful preparation of the physical system. This fact leads to some problematic issues for the concept of properties attached to observables. For example, imagine a two-dimensional vector space, and in it the observable σ_z with two eigenvectors: $\sigma_z|+\rangle = |+\rangle$, and $\sigma_z|-\rangle = -|-\rangle$. The observable σ_z thus has two possible outcomes for experiments: "+" or "−." If the system is prepared initially in the state $|+\rangle$, a measurement of σ_z will always yield the outcome "+", even if repeated several times, and similarly for $|-\rangle$. However, what happens to the state $|\psi\rangle = c_+|+\rangle + c_-|-\rangle$, the superposition of $|+\rangle$ and $|-\rangle$, where $|c_+|^2 + |c_-|^2 = 1$, when we measure property σ_z? We will either observe "+" or "−", and if measured again, we will observe the same outcome. This means that when we measure σ_z the state $|\psi\rangle$ "collapses" into either $|+\rangle$ or $|-\rangle$ with probabilities $p(+) = |c_+|^2$ and $p(-) = |c_-|^2$; we never observe the property to actually be in a superposition.

So, it is natural to ask about the meaning of superpositions of vectors such as $|\psi\rangle = c_+|+\rangle + c_-|-\rangle$. Does it mean that we have either state $|+\rangle$ or $|-\rangle$ but we do not know which? Or does it mean something different? It so happens that the idea that a superposition is a state with either one property or the other is not consistent.[1] So, a measurement does not reveal the existing value of a property, but seems to create it.

The puzzling aspect of quantum superpositions is perhaps better exemplified by the famous cat gedankenexperiment proposed by Schroedinger. In this experiment, a system has in it a cat. The cat can be described by two possible states, $|\text{livecat}\rangle$ and $|\text{deadcat}\rangle$, corresponding to the cat being either alive in one and dead in the other (i.e., repeated measurements of the cat lead to the same conclusions). Then, through a simple mechanism that constructs quantum superpositions for these two states, Schroedinger argued that it is possible to reach a superposed state of the type $c_l|\text{livecat}\rangle + c_d|\text{deadcat}\rangle$, such that for this state we cannot say that the observation reveals that the cat was either dead or alive. Instead, before the observation the cat

[1] This result is the core of the Kochen-Specker theorem, but showing it here would go beyond the scope of this paper. Interested readers are referred to the original paper of Kochen and Specker (1967).

is neither (or both), and the observation itself causes the cat to fall into the category of either dead or alive.

At the core of the seemingly absurd example from Schroedinger's cat seems to be the issue of observations. What are observations? What is to observe? To model observations, von Neumann (1983) proposed the following idea. Observations, he said, require something that determines which properties are being observed: call it a measurement apparatus. For example, if we want to observe if the cat is dead or alive, we can construct an apparatus that measures the cat's heartbeat: if a heartbeat is present, the cat is alive; otherwise, the cat is dead. Measuring apparatuses are physical systems themselves, and as such can be described by quantum physics. Now, quantum physics says that a system described by a vector $|\psi\rangle$ evolves according to a unitary evolution, i.e. $|\psi(T)\rangle = U(T, T_0)|\psi(T_0)\rangle$, where $|\psi(t)\rangle$ is the state at time t and $U(T, T_0)$ is the evolution operator.

The measurement apparatus modeling done by von Neumann does not resolve the problem of the cat superposition. This is because, during a measurement, the unitary evolution of the systems results in a superposition of the measurement apparatus with its state pointing to the cat dead and the other to the cat alive. So, von Neumann then argued the following. Since the measurement apparatus is in a superposition, then we can measure it with our eyes (themselves a measurement apparatus); this leads to a superposition of the dead and alive images in our retinas. But then we can consider the optic nerves as measuring the eyes themselves, and the optical nerves would be in a superposition. This would keep going on, until all matter in the brain/body would be in a quantum superposition of states where the cat dead or alive were imprinted. However, we never see such superposition. This led von Neumann to postulate two different dynamics for quantum processes: one, linear, unitary, and deterministic, that tells us how matter evolves in time; another, non-linear and probabilistic, which comes into play only when an observer experiences the measurement.

The idea proposed by von Neumann was later on modified, first by London and Bauer, and then by Wigner (see Bueno (2019)) in this volume for details and references), to include the mind. They argued that if all matter followed unitary and deterministic dynamics, then the mind had to be something else, something that was not matter. This is the idea behind the Consciousness Causes Collapse interpretation: matter evolves differently when it interacts with the conscious mind, which itself is not matter. This idea was later on abandoned by Wigner, but Henry Stapp continued to be one of its main proponents.

We can see how the CCC interpretation solves the measurement problem, and many readers of this book can imagine the many reasons why such proposal is not widely appreciated by physicists. But what are the alternatives? What other possibilities exist for the measurement problem. It would be beyond the scope of this paper to detail all the existing interpretations, and our goal here is simply to briefly comment on some popular interpretations, and what are the problems with them.

Let us start with hidden-variable theories. The idea of hidden variables is that quantum mechanics is an incomplete theory, and it is the goal of physicists to find a more complete description of nature. Perhaps this new theory includes unobservable properties (aptly called hidden variables) that can account for the quantum observed phenomena, and can be used to solve the measurement problem.

Perhaps the best known hidden-variable theory is Bohm's pilot-wave interpretation. In Bohm's theory, a quantum system is represented by a wave function $\psi(t, \mathbf{r})$, a vector in an infinite dimensional vector space, *and* a particle that "rides" this wave. This dynamics is described by a modified Hamilton-Jacobi equation with an added quantum potential. In this interpretation, the wave function has a dual aspect to it: it is both epistemic, as $|\psi(t, \mathbf{r})|^2$ gives us the probability of where the particle is at time t due to our lack of knowledge about it, and it is ontic, as its presence physically guides the particle. The main issue, according to many physicists, is the explicit non-locality of this interpretation. When two particles are described by a wave function of the form $\psi(t, \mathbf{r}_1, \mathbf{r}_2)$, where the index refers to either particle 1 or 2, an action (e.g. a measurement) in one of those particles would instantly affect the wave acting on the other particle, and since the wave has this ontic character, this amounts to an instantaneous action-at-a-distance, and Bohm's theory is non-local. This non-locality is seem by many as a major incompatibility with relativity theory (see Holland (1995) for a more detailed discussion of Bohm's theory and its criticisms).[2]

Another approach to the measurement problem is to deny the basic idea that a measurement yields a single outcome. For example, for Schroedinger's cat, the assumption is that when we observe the cat, we either get a dead cat or an alive cat. But a possibility is that both dead and alive cats are something that happen. To reconcile this with our experience, the many-worlds interpretation (Everett III 1957) assumes that in the process of a measurement, its possible outcomes (i.e. dead or alive cat) all happen, but in different universes. There are many reasons why people are reluctant to accept this theory, but perhaps the main difficulty is its very high ontological cost, where an infinitely large number of co-existing universes may pop up into existence at every instant. Despite this, the many-worlds (and its variant, the many-minds) interpretation is, perhaps, the most popular among physicists working on foundations (Schlosshauer et al. 2013).

A third approach to the measurement problem is to deny the universality of quantum unitary evolution. This is the approach of the CCC interpretation, as it says that, e.g., Schroedinger's equation applies to all matter, but not to mind. Furthermore, it postulates that the interaction of mind with matter yields a different type of dynamics, a non-unitary one. Since this is the main approach we are interested, we will discuss it in more detail.

Let S be a quantum system, represented by a vector that can take two states, $|+\rangle$ and $|-\rangle$. Let us assume that M is a measurement apparatus that has its own states described by $|\text{Pointer at } 0\rangle$, $|\text{Pointer at } +\rangle$, and $|\text{Pointer at } -\rangle$, where each of

[2]Though the author of this paper does not consider himself a "Bohmian," he believes that theories such as Bohm's are extremely important to get a fuller picture of what nature is trying to tell us. For example, it can be argued that Bohm's interpretation can lead to different outcomes from other interpretations when applied to extreme conditions, such as when the universe as a whole was dictated by quantum effects (de Barros and Pinto-Neto 1997, 1998; de Barros et al. 1998, 2000).

those states may represent a degenerate quantum state that has a pointer indicating that the system it measured has property + or −, or that it is simply waiting for a measurement. If M measures the properties of the state, its interaction with S should satisfy the following:

$$|+\rangle|\text{Pointer at }0\rangle \xrightarrow[\text{interaction}]{\text{S and }M} |+\rangle|\text{Pointer at }+\rangle, \quad (5.1)$$

$$|-\rangle|\text{Pointer at }0\rangle \xrightarrow[\text{interaction}]{\text{S and }M} |-\rangle|\text{Pointer at }-\rangle. \quad (5.2)$$

In terms of the quantum formalism, this means that there should be a unitary operator U_{measure} that when applied to the left hand side of (5.1) and (5.2) yields their right hand side, i.e.

$$U_{\text{measure}}|+\rangle|\text{Pointer at }0\rangle = |+\rangle|\text{Pointer at }+\rangle$$

and

$$U_{\text{measure}}|-\rangle|\text{Pointer at }0\rangle = |+\rangle|\text{Pointer at }-\rangle.$$

What happens when a measuring device, modeled by the quantum unitary evolution, interacts with a superposition? Let S be in the state $c_+|+\rangle + c_-|-\rangle$. Then, the evolution of S and M during their interaction is determined by U_{measure}, and after a measurement the final state is

$$U_{\text{measure}}\left(c_+|+\rangle + c_-|-\rangle\right)|\text{Pointer at }0\rangle = c_+|+\rangle|\text{Pointer at }+\rangle + c_-|-\rangle|\text{Pointer at }-\rangle. \quad (5.3)$$

A careful analysis of (5.3) reveals that, after a measurement, the final state is neither $|+\rangle|\text{Pointer at }+\rangle$ nor $|-\rangle|\text{Pointer at }-\rangle$, but instead a superposition of these two states. The consequence is that the final state is not an eigenstate of either $|+\rangle\langle+|$ or $|\text{Pointer at }+\rangle\langle\text{Pointer at }+|$, which are the projection observables associate to measurements of those states. In other words, no collapse of the wave function happened, and therefore no measurement happened.

As mentioned above, von Neumann (1983) argued that we could think of our eyes (call them system E) as a measurement apparatus that measures M itself. If we do that, and describe our eyes' interactions with S and M via an unitary operator, we end up with a similar superposition to (5.3). We could go further, and think of the optic nerve O as measuring the eyes, and the visual cortex V as measuring O, the brain B measuring V, and so on, until all matter in the body (and outside of it) is taken into account as part of the description, and we still end up with a superposition. But since we never see a superposition, the *observer* causes a different dynamics in the process that is not unitary: the collapse of the wave function.

5.3 Mind, Brain, and the IQZE

The above description is, briefly, the CCC interpretation. As we mentioned, this is a very controversial interpretation, and many researchers have not only argued against it in philosophical grounds, but claimed that it has been falsified. However, claims that the CCC interpretation was falsified have been greatly exaggerated. For instance, in de Barros and Oas (2017) it was argued that in order to falsify the CCC interpretation, one would need to perform an interference-type experiment with an organism candidate for having consciousness.[3] However, to prepare such an experiment, one would need to control all quantum variables that could get entangled with the measurement apparatus. This task is not solely difficult to realize from a technical point of view, but it would require keeping the organism decoupled from a thermal bath, at temperatures that are incompatible neither with life nor with behavioral responses, not to mention consciousness. But even though CCC probably will never be falsified, at least not in a way that is convincing to many of its proponents, it does not mean that it cannot raise some interesting, and falsifiable, models of interaction of mind and brain. This is what we will explore in this section.

Here we will focus on Stapp's work, in particular his proposal of how the mind affects matter in particular ways, e.g. how the non-material mind controls the motor cortex such that one's fingers type the words one wishes in a keyboard (Stapp 2009, 2014). Stapp's proposal relies on the Inverse Quantum Zeno Effect (IQZE), which we describe below. The idea of the Quantum Zeno Effect is that it is possible to create systems and observables such that the expected values of some observables change with time when the system is left to evolve regularly but remain constant when the system is being observed through time. The IQZE is the opposite: the observation of a system whose observables would remain the same leads to changes in time for the value of those observables. To show how this works, imagine a quantum harmonic oscillator represented by a coherent state $|\alpha\rangle$, defined by

$$|\alpha\rangle = e^{-|\alpha|^2/2} \sum \frac{\alpha^n}{\sqrt{n!}} |n\rangle,$$

where α can be interpreted as the amplitude of oscillation. Imagine that we have as our observer the projection operator

$$P_{\alpha+\epsilon} = |\alpha + \epsilon\rangle\langle\alpha + \epsilon|,$$

[3] Here we should clarify what we mean by consciousness. von Neumann seemed to be thinking about a generic concept, that of an observer. He did not (at least explicitly) commit to a mind or consciousness that was outside of the physical realm. This connection between a nonphysical mind seem to be espoused initially by Wigner, and later on by Henry Stapp. But even for Stapp's view, it seems that what is meant by consciousness is not phenomenal consciousness, but access consciousness (see Montemayor (2019) on this volume).

which corresponds to the question "is the system in a coherent state with amplitude $\alpha + \epsilon$?" If we perform such measurement for the coherent state $|\alpha\rangle$, we have two possible outcomes: the system ends in a coherent stat $|\alpha + \epsilon\rangle$ with probability $p(\alpha + \epsilon|\alpha) = 1 - \epsilon^2$; the system does not have the property $\alpha + \epsilon$ with probability ϵ^2. Now, let us assume that we make N measurements like the above one, but in the following order: $P_{\alpha+\epsilon}$, $P_{\alpha+2\epsilon}$, $P_{\alpha+3\epsilon}, \ldots, P_{\alpha+N\epsilon}$. It follows, because of the collapse of the wave function for each observation, that after these N measurements the system will be in the state $|\alpha + N\epsilon\rangle$ with probability $p(\alpha + N\epsilon) \approx 1 - N\epsilon^2$, for small values of ϵ. If we take the limit for $\epsilon \to 0$ while keeping $N\epsilon = \beta$ constant, the system ends up in the state $|\alpha + \beta\rangle$ with probability one. In other words, the continuously increasing measurement of quantum oscillator can lead to its change of amplitude.

Notice that the IQZE happens because we have a collapse of the wave function. If there were no collapse, a multitude of superpositions would be created, and the final state would not be $|\alpha + \beta\rangle$ with probability one. So, keeping this in mind, Stapp proposed to use the IQZE as a way that the mind could influence matter. According to Stapp, small neural oscillators on the cortex micropill could be measured in a similar way as the IQZE, and what was originally a small oscillation could become a very large one (in fact, as large as one wishes): if the mind chooses to observe the oscillator in a particular way, the measurement causes an increase in the amplitude of neural oscillators; if the mind does not choose to observe, there is no increase in the amplitude of neural oscillators.

5.4 Where Mind Meets Matter

In previous work, we argued that the model presented by Stapp is incomplete (de Barros 2016; de Barros and Oas 2015). Our basic reasoning was, briefly, the following. Even in the CCC interpretation, a measurement does not constitute an interaction between a system and the mind. Instead it is the interaction of a system with a measurement apparatus and with a mind. The measurement apparatus is a crucial component of the measurement, as it defines the measurement basis (Schlosshauer 2005). In other words, what determines whether we are measuring spin in the direction z is not the mind of the observer, but the fact that our measurement apparatus is physically placed in a way that measures spin in that direction. If we were to physically rotate this measurement apparatus to the direction y, we would obtain measurements in the direction y, regardless of whether our mind wanted or not.

So, here we seem to hit a apparent problem: for mind to affect matter, we need to have a measurement apparatus that moves in a very particular way. But this measurement apparatus needs to be moved by mind. So, this seems to lead to a circular argument, one where mind affects matter through measurements (via the IQZE) only if mind can physically affect matter (through the positioning of measurement apparatuses). Does this mean that we need to abandon the IQZE model? Not necessarily. As we showed before (de Barros and Oas 2015), it is possible to resolve this issue of circularity if we make additional assumptions to the dynamics of the brain. Furthermore, those additional assumptions perhaps may provide us with specific types of structures that may be involved in the interaction between mind and matter, so let us examine them now.

First, the issue of a measurement and its basis is essential for a measurement. This is, after all, the main success of decoherence theory: the explanation of why a measurement yields a preferred basis. So, any theory of measurement, which CCC is, needs to include this. In other words, the first constraint on Stapp's model is that, in addition to the neural oscillators in a coherent state $|\alpha\rangle$, there needs to be a physical structure that provides a preferred basis for the successive measurements of $P_{\alpha+\epsilon}, P_{\alpha+2\epsilon}, \ldots, P_{\alpha+N\epsilon}$. Let us call this the *brain-mind mediator structure* (BMMS), as it mediates the interaction between mind and matter.

In addition to interacting with the neuropill oscillators (NO) as a measurement device with increasing amplitude, BMMS has to offer some predictability, for the following reason. If the mind is to control MP, choosing freely to increase or not its amplitudes, it has two choices: observe the BMMS or not. Note that for a measurement to occurs, according to the CCC theory, we need a physical system being measured (here the NO), a physical measurement apparatus (BMMS), and an observer's mind (OM). If OM observes, then a collapse occurs, and the amplitude of $|\alpha\rangle$ increases. If OM does not observe, the amplitude of $|\alpha\rangle$ remains the same. But given that the observation needs to start happening when BMMS is at the state of measuring $P_{\alpha+\epsilon}$, and not later, and then keep measuring until the desired amplitude is reached, the dynamics of the measurements need to be predictable, and it needs to be such that the mind can choose the measurement as often as it wishes. This can be accomplished by assuming that BMMS's dynamics is periodic, but how this periodicity is described (i.e. what are the exact details of this oscillations) is not clear.

So, here we reach the constraints involved in the brain structures needed for Stapp's model to work.

- NO. The neuropill oscillators are the basic structures that are associated with the increase of activity due to the mind's volition. Stapp assumes them to be in a quantum coherent state, $|\alpha\rangle$. Though this is a simplifying assumption, there are other quantum oscillation states that could be increased by a similar effect to the IQZE. But, effectively, what we would need to be looking for in the brain are structures that have some type of oscillation that could be directly linked to a behavioral response.
- BMSS. The NO needs to be interacting with BMSS to be a candidate for where the mind meets the brain. BMMS needs to have some sort of predictability, i.e. it needs to be periodic (though its specific dynamics may be very complicated). Furthermore, BMMS interaction with NO has to provide a preferred basis of measurement that allows for the increase in activity of the NO if a mind observes it.

The neural system NO+BMMS constitute the locus on which mind interacts with matter in a CCC model consistent with Stapp's.

5.5 Conclusions

In this paper we examined Stapp's model of interaction of the mind with brain using the CCC interpretation of quantum mechanics. We discussed some constraints associated with Stapp's model, and argued that some very specific dynamics for brain structures need to emerge, if the mind were to affect the brain structures according to its volition. These constraints, together with the dynamics, should provide a way for us to investigate the existence of structures that allow for the interaction of the mind and brain, and would be indicative of Stapp's model.

We should emphasize that a major assumption was made in this paper, which is that the mind interacts with matter in the brain. It is quite possible that the mind actually does not interact at all in the brain, or not even in the structures that we identified as NO and BMMS. Such structures could, in principle, get entangled with other environmental components, and the interaction of the mind with those other components would lead to the same type of outcome: a control of matter from a nonphysical mind. However, the interaction with other components would have to be done in a way that could keep the independence of each BMMS associated to distinct NOs and diverse behavioral outcomes. I.e., the NO associate with moving one's right arm would need to be isolated, from a quantum perspective, to the oscillators associated to moving one's left arm. So, this suggests that the mind observations happen at a very local level, and clearly at an unconscious way.

Another issue that is important to mention is related to Wigner's friend. If we were to identify the structures in question, and then allow them to be observed also by an external observer (the experimenter), it would be possible, in principle, that this external observer would have an effect. For example, let us say that we can pinpoint the NO+BMMS associated with rising a participant's right leg. If we were to observe this oscillator, with no physical interaction with it other than, perhaps, a weak interaction that would entangle its quantum degrees of freedom with our mind, we would be able to make the participant's legs rise. In other words, even though the participant did not want to rise their legs, the observation by an external mind caused them to do so.

Finally, the last issue that should be addressed is about the NO+BMMS systems that were not observed. How are those systems prevented from being entangled with environmental variables, such that they are always being measured by the observer, or even by an external observer? Or are there other structures that are observed, and that prevent the quantum coherence to proceed and be affected by external observers?

Though the CCC interpretation is not taken seriously by many physicists, it does solve the measurement problem. And even though one could argue that it cannot be falsified, it does provide some interesting opportunities to think about modeling the mind interacting with matter, e.g. with Stapp's model. These models may provide constraints to brain structures that could be, in principle, observed. Furthermore, as we saw, this types of model raise many important, and falsifiable, questions, the mark of an interesting scientific theory. It is the hope of this article to provide some grounds for these types of discussions.

References

Bohm, D. (1952a). A suggested interpretation of the quantum theory in terms of "Hidden" variables. I. *Physical Review, 85*(2), 166–179.

Bohm, D. (1952b). A suggested interpretation of the quantum theory in terms of "Hidden" variables. II. *Physical Review, 85*(2), 180–193.

Bueno, O. (2019). Is there a place for consciousness in quantum mechanics? In J. A. de Barros & C. Montemayor (Eds.), *Quanta and mind: Essays on the connection between quantum theories and consciousness* (Synthese library).

de Barros, J. A. (2016). On a model of quantum mechanics and the mind. In S. O'Nuallain (Ed.), *arXiv:1404.0714 [quant-ph, q-bio]*. Berkeley: Cambridge Scholars Publishing.

de Barros, J. A., & Oas, G. (2015). Quantum mechanics & the brain, and some of its consequences. *Cosmos and History: The Journal of Natural and Social Philosophy, 11*(2), 146–153.

de Barros, J. A., & Oas, G. (2017). Can we falsify the consciousness-causes-collapse hypothesis in quantum mechanics? *Foundations of Physics, 47*(10), 1294–1308.

de Barros, J. A., & Pinto-Neto, N. (1997). The causal interpretation of quantum mechanics and the singularity problem in quantum cosmology. *Nuclear Physics B-Proceedings Supplements, 57*(1–3), 247–250.

de Barros, J. A., & Pinto-Neto, N. (1998). The causal interpretation of quantum mechanics and the singularity problem and time issue in quantum cosmology. *International Journal of Modern Physics D, 7*(02), 201–213.

de Barros, J. A., Pinto-Neto, N., & Sagioro-Leal, M. A. (1998). The causal interpretation of dust and radiation fluid non-singular quantum cosmologies. *Physics Letters A, 241*(4), 229–239.

de Barros, J. A., Pinto-Neto, N., & Sagioro-Leal, M. A. (2000). The causal interpretation of conformally coupled scalar field quantum cosmology. *General Relativity and Gravitation, 32*(1), 15–39.

Everett III, H. (1957). "Relative State" formulation of quantum mechanics. *Reviews of Modern Physics, 29*(3), 454–462.

Fuchs, C. A. (2014). Introducing QBism – Springer. In M. Galavotti, D. Dieks, J. Gonzales, S. Hartmann, T. Uebel, & M. Weber (Eds.), *The philosophy of science in a European perspective*. New directions in the philosophy of science. Dordrecht/Heidelberg: Springer.

Holland, P. R. (1995). *The quantum theory of motion: An account of the de Broglie-Bohm causal interpretation of quantum mechanics*. Cambridge: Cambridge University Press.

Kochen, S., & Specker, E. P. (1967). The problem of hidden variables in quantum mechanics. *Journal of Mathematics and Mechanics, 17*, 59–87.

Montemayor, C. (2019). Panpsychism and quantum mechanics: Explanatory challenges. In J. A. de Barros & C. Montemayor (Eds.), *Quanta and mind: Essays on the connection between quantum theories and consciousness* (Synthese library).

Nelson, E. (1985). *Quantum fluctuations*. Princeton: Princeton University Press.

Schlosshauer, M. (2005). Decoherence, the measurement problem, and interpretations of quantum mechanics. *Reviews of Modern Physics, 76*(4), 1267–1305.

Schlosshauer, M., Kofler, J., & Zeilinger, A. (2013). A snapshot of foundational attitudes toward quantum mechanics. *Studies in History and Philosophy of Science Part B: Studies in History and Philosophy of Modern Physics, 44*(3), 222–230.

Stapp, H. P. (2009). Mind, matter, and quantum mechanics. In H. P. Stapp (Ed.), *Mind, matter and quantum mechanics*, the Frontiers collection (pp. 81–118). Berlin/Heidelberg: Springer.
Stapp, H. P. (2014). Mind, brain, and neuroscience. *Cosmos and History, 10*(1), 227–231.
von Neumann, J. (1983). *Mathematical foundations of quantum mechanics*. Princeton: Princeton University Press. Translated by R. T. Beyer from the 1932 german edition.

Chapter 6
Quantum Schmuntum?

Paweł Kurzyński and Dagomir Kaszlikowski

Abstract Although macroscopic objects, from simple gases to highly complex biological matter, are made of a large number of microscopic structures that can be in principle described by quantum physics, we argue that it is not enough to assume that they cannot be efficiently described by classical theory. We speculate that almost all of them can be as efficiently simulated on classical computers as on quantum ones, or that their quantum simulations are as inefficient as classical simulations.

6.1 Introduction

It is certain that everything around us, from simple rocks to complex human brains, is made of microscopic constituents that obey quantum mechanical laws. It is also somewhat certain that it is not easy to observe quantum phenomena in the macroscopic (classical) world. In most cases the fast quantum-to-classical transitions are attributed to decoherence (Zurek 1991). Decoherence happens in any open system and macroscopic objects certainly fall into this category. However, recent studies show that coherence times in realistic environments can be quite impressive (see for example Panitchayangkoon et al. 2010). Nevertheless, observation of coherence is not enough to claim that the given system is quantum. Coherence certainly implies a wave-like behaviour but not necessarily a quantum one.

What is quantumness then? Common knowledge seems to equate quantumness with the wave-particle duality: observation forces quantum systems into wave-like

P. Kurzyński (✉)
Faculty of Physics, Adam Mickiewicz University, Poznań, Poland
e-mail: pawel.kurzynski@amu.edu.pl

D. Kaszlikowski
Centre for Quantum Technologies, National University of Singapore, Singapore, Singapore

Department of Physics, National University of Singapore, Singapore, Singapore
e-mail: phykd@nus.edu.sg

J. Acacio de Barros, C. Montemayor (eds.), *Quanta and Mind*, Synthese Library 414, https://doi.org/10.1007/978-3-030-21908-6_6

or particle-like behaviour. This fundamental, if not paranoid, property of Nature leads to the phenomenon of complementarity, which states that some physical properties cannot be simultaneously observed. In addition, quantum theory only provides us with a recipe to estimate probabilities of observable measurement outcomes. For two complementary measurements, A and B, quantum mechanics gives a rule to estimate probability distributions $p(A = a)$ and $p(B = b)$ but says nothing about the joint probability distribution (JPD) $p(A = a, B = b)$. Hence, the JPD problem is one of the facets of complementarity. Still, even for complementary properties, it is often possible to construct a JPD for A and B that correctly reproduces measurable marginal probabilities $p(A = a)$, $p(B = b)$. In this case the system can be simulated by classical, deterministic parameters, usually referred to in literature as non-contextual hidden variables (NCHV) that encode all possible outcomes a, b. In such a scenario the system might be fundamentally quantum but it has an alternative, classical description. Thus to certify a truly quantum behaviour, one needs to show that it is impossible to construct a JPD compatible with measurable marginal probability distributions (Fine 1982). This is not always possible and Nature is provably quantum (Bell 1964; Kochen and Specker 1967).

In this work we argue that even in perfectly isolated conditions with no decoherence, natural properties of any system with only a few quatum particles can be quite easily simulated by a JPD. It is therefore extremely hard to have a truly quantum behaviour in systems consisting of more than one particle. Thus, we speculate that simple and complex macroscopic objects (such as living matter) can be as efficiently simulated on a classical computer as on a quantum one, or that their quantum simulations are as inefficient as classical simulations.

6.2 Quantumness and Joint Probability Distributions

Quantum theory is effectively an algorithm to calculate probabilities of measurement outcomes that can be performed on a quantum system. The system is prepared by an experimenter (sometimes by Nature itself) who also sets up a measuring apparatus that generates measurement outcomes. The outcomes are related to the elementary questions 'yes' (encoded as 1) and 'no' (encoded as 0) that the experimenter is allowed to ask. 0s and 1s seem to happen randomly and the quantum theory can precisely estimate how (im-)probable they are. Some questions cannot be jointly asked according to the quantum theory – They are usually called 'complementary' questions. For instance: 'is an atom at position x?' and 'is momentum of an atom p?' are complementary questions. Operationally, no one has ever constructed a measuring apparatus that can perform *sharp*, i.e., 'yes' and 'no' measurements of complementary observables on an arbitrary quantum state.

However, this seemingly fundamental limitation may come from some sort of incompleteness of quantum theory as famously suggested by Einstein, Podolsky and Rosen (EPR) in their influential paper from mid 1930s (Einstein et al. 1935).

Indeed, there could be some hidden (from us) physical parameters, knowledge of which, would allows us to get definitive answers to complementary questions. These hypothetical parameters are commonly referred to as 'non-contextual hidden variables' (NCHV).

NCHV can be simply described as follows. Suppose, you want to perform measurements of some observables (complementary or not) A_i ($i = 1, 2, \ldots, N$) with k-valued outcomes a_i. NCHV λ_l encode all possible measurement outcomes that A_i can generate, i.e., $\lambda_l = (a_1, a_2, \ldots, a_N)$ with $l = 1, 2, \ldots, k^N$. To account for randomness one imposes some probability distribution (JPD) p on λ_l, i.e., $p(\lambda_l) = p(a_1, a_2, \ldots, a_N)$. The only requirement for $p(\lambda_l)$ is that it agrees with experimental observations: marginals of $p(\lambda_l)$ must recover statistics of all allowed measurements. For instance, suppose you have four binary observables A_1, A_2 and B_1, B_2, complementary within A (B) group and jointly measurable across the groups. Experimentally available probabilities are $p(a_1, b_1)$, $p(a_1, b_2)$, $p(a_2, b_1)$, $p(a_2, b_2)$ and any JPD must recover them as its marginals: $\sum_{a_2,b_2} p(a_1, a_2; b_1, b_2) = p(a_1, b_1)$ and so on. Of course now, one can assign well defined probabilities to 'forbidden' complementary observables: $p(a_1, a_2) = \sum_{b_1,b_2} p(a_1, a_2; b_1, b_2)$ and $p(b_1, b_2) = \sum_{a_1,a_2} p(a_1, a_2; b_1, b_2)$. If they existed, NCHV would offer ultimate knowledge about any physical observables just like classical physics promised before the quantum theory was discovered.

Theoretical proofs of non-existence of NCHV were given independently by Bell in (1964) and Kochen and Specker in (1967). Experiments confirming theoretical predictions followed (Freedman and Clauser 1972; Aspect et al. 1982). It is not the scope of this paper to discuss all the intricacies of theoretical proofs and that of laboratory experiments but it is harmless to write a few words of explanation. The simplest to describe is Bell's proof who found an algebraic inequality (Bell inequality) for the experimentally available probabilities that must be obeyed if these probabilities are the marginals of NCHV's JPD. Violations of Bell's inequality indicate non-existence of NCHV.

To recapitulate, NCHV state that the outcomes of all possible experiments one can perform on a given physical system have definite, objective values. Those values are merely 'revealed' by the act of observation (measurement). It is clear that our macroscopic world can be successfully described by NCHV – I do not need to see directly the money on my bank account, it is enough to check my account statement on a mobile. Although NCHV do not exist, they capture the paradigm of classicality in the most general, experimentally verifiable way. As far as the authors of this paper are concerned, it is the only unambiguous test of classicality available to us.

A necessary condition for the lack of NCHV, and as a consequence the lack of a JPD, is to have a subset of observables that cyclically commute (Fine 1982). Here we are going to discuss one of the simplest scenarios consisting of the already mentioned set of four binary observables: A_1, A_2, B_1 and B_2. For the purpose of the scenario we assume that these observables have outcomes ± 1. The cyclic commutation is provided by $[A_i, B_j] = 0$ for $i, j = 1, 2$, but $[A_1, A_2] \neq 0$ and $[B_1, B_2] \neq 0$, so a JPD $p(a_1, a_2, b_1, b_2)$ cannot be directly measured. These four observables constitute the well known Clauser-Horne-Shimony-Holt (CHSH) scenario (Clauser et al. 1969).

The crucial observation is that if there is a JPD for the CHSH scenario, the following inequality must be satisfied

$$-2 \leq \langle A_1 B_1 \rangle + \langle A_1 B_2 \rangle + \langle A_2 B_1 \rangle - \langle A_2 B_2 \rangle \leq 2. \tag{6.1}$$

The above can be also expressed as

$$p(A_1 = B_1) + p(A_1 = B_2) + p(A_2 = B_1) + p(A_2 \neq B_2) \leq 3. \tag{6.2}$$

This inequality stems from the following chain of implications: if $A_2 = B_1$ and $B_1 = A_1$ and $A_1 = B_2$, then $A_2 = B_2$. In general, it is easy to check that if three of the above four probabilities are 1, then the last one is necessarily 0. This is implied by the outcomes' exclusivity.

The CHSH scenario can be realised in two settings. The fist setting is nonlocal, in which case the system is bipartite and observables A_1 and A_2 are measured on the first part, whereas B_1 and B_2 are measured on the second part. In nonlocal setting the violation of (6.1) and (6.2) can be observed only if the system is entangled. The second setting is local, in which case all four observables are measured on a local system that needs to have at least four distinct states.

It is known that quantum theory allows to violate the above inequalities, both in nonlocal and local settings. The inequality (6.1) can be violated up to $\pm 2\sqrt{2}$, whereas the inequality (6.2) can be violated up to $2+\sqrt{2}$. The violation of the CHSH inequality confirms the lack of NCHV and the corresponding JPD confirming the system's genuine quantumness. However, to have it one needs to prepare the system in a specific state and perform special measurements that address this particular system.

6.3 Classicality of Many Copies

Under realistic conditions it is extremely hard to measure individual properties of a single quantum system interacting with other quantum systems. Instead, one usually addresses whole ensembles of quantum systems. For example, in a bulk material or gas one does not address a magnetization of a single spin, but rather a collective magnetization of many spins. Such collective properties of N quantum systems q_1, $q_2, \ldots, q_N$ result from the measurements of the same physical property, say A, on each individual system. More precisely an observable $A^{(i)} = \mathbb{1}^{\otimes^{i-1}} \otimes A \otimes \mathbb{1}^{\otimes^{N-i+1}}$ is measured on the system q_i.

Let us now discuss the properties of such collective observables. Assume that the state of all N quantum systems is ρ, and that $\rho^{(i)}$ is a reduced state of the system q_i. It is clear, that the result of collective measurement does not change under the permutation of systems. More precisely, one would obtain the same collective measurement results if the measurement $A^{(i)}$ were performed on the system in the state $\rho^{(\pi_i)}$, where π_i is an arbitrary permutation of indices. We can therefore apply all possible permutations to symmetrize the N-partite state ρ, i.e., $\rho \to \rho_{sym}$ without any observational difference. Moreover, for ρ_{sym} every reduced state is the same, i.e., $\rho^{(i)} = \sigma$. Therefore, from the point of view of collective measurements all N systems are indistinguishable.

Next, we ask if we can find NCHV for collective measurements. This problem was addressed by us and our collaborators in a collection of papers (Ramanathan et al. 2011; Kurzynski et al. 2013; Kurzynski and Kaszlikowski 2016; Markiewicz et al. 2019). We refer the reader to these works for more details. Here we focus on the CHSH scenario and show that even for $N = 2$ there exists a classical description.

Let us first focus on the standard scenario with $N = 1$ and the binary ± 1 observables A_1, A_2, B_1 and B_2. The inequalities (6.1) and (6.2) can be expressed using the probabilities corresponding to the following eight events:

$$(++|11), (--|11), (++|12), (--|12), (++|21), (--|21), (+-|22), (-+|22),$$

where $(ij|kl)$ means that A_k and B_l were measured and the observed results were i and j, respectively. These events lead to the following 12 pairs of mutually exclusive events (Cabello 2013)

$$\{(++|11), (--|11)\}, \{(++|12), (--|12)\}, \{(++|21), (--|21)\}, \{(+-|22), (-+|22)\},$$

$$\{(++|11), (--|12)\}, \{(-+|22), (++|21)\}, \{(--|11), (++|12)\}, \{(+-|22), (--|21)\},$$

$$\{(++|11), (--|21)\}, \{(+-|22), (++|12)\}, \{(--|11), (++|21)\}, \{(-+|22), (--|12)\}.$$

The pairs in the second and the third row are exclusive because they share exactly one measurement, but the corresponding outcomes are opposite ($+$ versus $-$). If we try to construct an NCHV model by assigning deterministic values to these eight events, 1 if event happens and 0 if it does not happen, we need to remember that for exclusive events at most one event can be assigned 1. Moreover, the noncontextuality assumption states that if we assign some value to an event, this value must be the same, no matter with which other event it is measured (Kochen and Specker 1967). It is easy to check that the exclusivity structure presented above implies that at most three events can be assigned the value 1, hence the upper bound of (6.2) is three. However, if we find in an experiments that the sum of four measurable probabilities is greater than three, then we know that there is no NCHV description.

Next, we consider the above scenario with $N = 2$. This time when we measure A_k and B_l we have more than four exclusive outcomes: $(++|kl)$, $(+-|kl)$, $(-+|kl)$ and $(- - |kl)$. Each of two systems can produce a different pair of outcomes, therefore in principle there are 16 possible exclusive events. However, since we argued that we are limited to collective observables, which do not distinguish between the systems, we have only 10 possible exclusive events:

$$(++,++|kl), (++,+-|kl), (++,-+|kl), (++,--|kl), (+-,+-|kl),$$

$$(+-,-+|kl), (+-,--|kl), (-+,-+|kl), (-+,--|kl), (--,--|kl).$$

The notation $(ij,i'j'|kl)$ means that A_k and B_l were measured on both systems and one system produced the outcomes i and j, whereas the other produced the outcomes i' and j'.

There are two particularly interesting events in the above set. The first one is $(++,--|kl)$, which implies that A_k and B_l are perfectly correlated. The other one is $(+-,-+|kl)$, which implies that these observables are perfectly anti-correlated. Now, consider the NCHV model in which we assign the deterministic value 1 to the following four events:

$$(++,--|11),\ (++,--|12),\ (++,--|21),\ (+-,-+|22).$$

How is that possible? The solution is the following – this time there is no exclusivity between the above events. For example, $(++,--|11)$ and $(++,--|12)$ are not exclusive because in each event for the common measurement A_1 one system produced the value $+$ and another produced the value $-$. The same is true for all the other pairs. Note, that the exclusivity would be present if the systems were distinguishable, because in this case we would need to specify which system produced which outcome which would be equivalent to considering the case $N=1$ twice.

Interestingly, the above NCHV model implies that collective measurements lead to $A_2=B_1$, $B_1=A_1$, $A_1=B_2$ and $A_2=-B_2$, which is not possible in case $N=1$. As a result, for $N=2$ the violation of (6.1) and (6.2) does not imply the lack of NCHV and therefore such violation does not imply any quantumness (this also explains the nature of violations studied for example in Suppes et al. 1996, Spreeuw 1998, Spreeuw 2001, Qian and Eberly 2011, Qian and Eberly 2013, Aiello et al. 2015, Qian et al. 2015, Snoke 2014, Frustaglia et al. 2016). Finally, note that the corresponding classical explanation has nothing to do with decoherence and works even if both systems are perfectly isolated from the outside world.

6.4 Conclusions

It is quite reasonable to assume that Nature is optimal. However, the notion of optimality is not absolute. For example, in the theory of computation we can define the optimality with respect to a number of operations needed to complete some task. We can also define the optimality with respect to a memory needed to store outcomes of intermediate computation steps, or a memory needed to encode an algorithm. The problem is that the optimality with respect to one resource does not imply the optimality with respect to some other resource.

It is possible that the working of many macroscopic complex objects, like the brain, is truly quantum. However, one must understand that while quantum theory offers some new effects, that do not occur in classical systems, one needs to pay a price to observe them. Namely, one needs to isolate the system from the environment and one needs to prepare it in a special state. Next, one needs to apply extremely precise operations and finally one needs to employ highly efficient detection techniques. Therefore, while quantum phenomena may offer an optimality with respect to some resources, the observation of these phenomena requires a large amount of other resources. This raises a practical question: is it worth it?

The above question is constantly addressed by scientists from the quantum information and computation community, and we believe that it is safe to say that they are still far from giving a definite answer. For example, the violation of the CHSH scenario in the nonlocal setting is provided by the correlations resulting from the entanglement between two two-level systems. However, a communication of a single classical bit from one party to the other is enough to classically simulate such correlations (Toner and Bacon 2003). In a similar spirit, correlations in local non-classical contextual systems can be simulated classically if one uses few additional bits of memory (Kleinmann et al. 2011; Fagundes and Kleinmann 2017). Although, the single bit of communication and the additional bit of memory is fundamentally important for our understanding of Nature, it does not mean that it is important for every practical purpose. Therefore, one may ask if it is not simpler and more efficient to use extra bits of communication or to provide few additional bits of memory than to perfectly isolate a single quantum system to individually address its properties with sophisticated machinery? We believe that in most situations occurring in Nature it is easier to go classical.

Especially, collective measurements, which were shown to lead to classical description, are quite natural in systems that should be highly resistant to noise. Noisy systems cannot rely on a succinct encoding of information. Instead, they have to use redundancies to protect information from any damage (Shannon 1948). In the simplest scenario one replicates information, so if one copy is destroyed, there is a high chance that the other ones will remain intact. This way of dealing with noise is used in many natural and engineered systems. The most important aspect of this approach is that noisy systems no longer rely on exact values of single bits, but rather use average values evaluated on many bits. For example, the activation of a logic gate or neuron is not caused by a single charge, but by an electric current of an intensity larger than some threshold value. Despite existing proposals of highly sophisticated quantum-error correcting codes Nielsen and Chuang (2000), we speculate that classical theory is as efficient as quantum theory in describing the working of majority of systems in Nature, even on the most fundamental level.

References

Aiello, A., Toppel, F., Marquardt, C., Giacobino, E., & Leuchs, G. (2015). Quantum-like nonseparable structures in optical beams. *New Journal of Physics, 17*, 043024.

Aspect, A., Dalibard, J., & Roger, G. (1982). Experimental test of Bell's inequalities using time-varying analyzers. *Physical Review Letters, 49*, 1804.

Bell, J. S. (1964). On the Einstein Podolsky Rosen paradox. *Physics (Long Island City, N.Y.), 1*, 195.

Cabello, A. (2013). Simple explanation of the quantum violation of a fundamental inequality. *Physical Review Letters, 110*, 060402.

Clauser, J. F., Horne, M. A., Shimony, A., & Holt, R. A. (1969). Proposed experiment to test local hidden-variable theories. *Physical Review Letters, 23*, 880.

Einstein, A., Podolsky, B., & Rosen, N. (1935). Can quantum-mechanical description of physical reality be considered complete? *Physics Review, 47*, 777.

Fagundes, G., & Kleinmann, M. (2017). Memory cost for simulating all quantum correlations from the Peres-Mermin scenario. *Journal of Physics A: Mathematical and Theoretical, 50*, 325302.

Fine, A. (1982). Hidden variables, joint probability, and the bell inequalities. *Physical Review Letters, 48*, 291.

Freedman, S. J., & Clauser, J. F. (1972). Experimental test of local hidden-variable theories. *Physical Review Letters, 28*, 938.

Frustaglia, D., Baltanas, J. P., Velazquez-Ahumada, M. C., Fernandez-Prieto, A., Lujambio, A., Losada, V., Freire, M. J., & Cabello, A. (2016). Classical physics and the bounds of quantum correlations. *Physical Review Letters, 116*, 250404.

Kleinmann, M., Guhne, O., Portillo, J. R., Larsson, J.-A., & Cabello, A. (2011). Memory cost of quantum contextuality. *New Journal of Physics, 13*, 113011.

Kochen, S., & Specker, E. P. (1967). The problem of hidden variables in quantum mechanics. *Journal of Mathematics and Mechanics, 17*, 59–87.

Kurzynski, P., & Kaszlikowski, D. (2016). Macroscopic limit of nonclassical correlations. *Physical Review A, 93*, 022125.

Kurzynski, P., Soeda, A., Ramanathan, R., Grudka, A., Thompson, J., & Kaszlikowski, D. (2013). On the problem of contextuality in macroscopic magnetization measurements. *Physics Letters A, 377*, 2856.

Markiewicz, M., Kaszlikowski, D., Kurzynski, P., & Wojcik, A. (2019). From contextuality of a single photon to realism of an electromagnetic wave. *npj Quantum Information, 5,* Article number: 5.

Nielsen, M. A., & Chuang, I. L. (2000). *Quantum computation and quantum information*. Cambridge: Cambridge University Press.

Panitchayangkoon, G., Hayes, D., Fransted, K. A., Caram, J. R., Harel, E., Wen, J., Blankenship, R. E., & Engel, G. S. (2010). Long-lived quantum coherence in photosynthetic complexes at physiological temperature. *PNAS, 107*, 12766.

Qian, X.-F., & Eberly, J. H. (2011). Entanglement and classical polarization states. *Optics Letters, 36*, 4110.

Qian, X.-F., & Eberly, J. H. (2013). Entanglement is sometimes enough. arXiv:1307.3772.

Qian, X., Little, B., Howell, J., & Eberly, J. (2015). Shifting the quantum-classical boundary: theory and experiment for statistically classical optical fields. *Optica, 2*, 611.

Ramanathan, R., Paterek, T., Kay, A., Kurzynski, P., & Kaszlikowski, D. (2011). Local realism of macroscopic correlations. *Physical Review Letters, 107*, 060405.

Shannon, C. E. (1948). A mathematical theory of communication. *Bell System Technical Journal, 27*, 379623.

Snoke, D. (2014). A macroscopic classical system with entanglement. arXiv:1406.7023.

Spreeuw, R. J. C. (1998). A classical analogy of entanglement. *Foundations of Physics, 28*, 361.

Spreeuw, R. J. C. (2001). Classical wave-optics analogy of quantum-information processing. *Physical Review A, 63*, 062302.

Suppes, P., de Barros, J. A., & Sant'Anna, A. S. (1996). A proposed experiment showing that classical fields can violate bell's inequalities. arXiv:quant-ph/9606019.

Toner, B. F., & Bacon, D. (2003). Communication cost of simulating bell correlations. *Physical Review Letters, 91*, 187904.

Zurek, W. H. (1991). Decoherence and the transition from quantum to classical. *Physics Today, 44*, 36–44.

Chapter 7
A Quantum Model of Non-illusory Free Will

Kathryn Blackmond Laskey

Abstract Contemporary science and philosophy are dominated by a mechanistic materialist metaphysic that treats consciousness as a derivative aspect of the brain's physical state, with no independent causal efficacy ascribed to consciousness. Studies suggest there may be negative social consequences to widespread popular belief that our thoughts are passive spectators to our behavior. Dissenting from the commonly held view, the psychologist William James argued that consciousness must serve some evolutionary purpose, and therefore must be efficacious. But how might something as insubstantial as a thought cause something to happen in the physical world? According to mechanistic materialism, it cannot. However, there is an alternative to mechanistic materialism. The physicist Stapp argues that a realistic interpretation of quantum theory can form the basis for a scientifically well-founded theory of efficacious conscious choice. The resulting theory of agency fills complementary explanatory gaps in physics and psychology, allowing consciousness to become efficacious in a manner entirely consistent with empirically validated physical theory. The profound implications of a scientifically well-founded theory of non-illusory free will argue for working out a detailed model of its operation in human brains and devising empirical tests of the model's predictions.

Keywords Free will · Quantum Zeno effect · Neural networks · Synchronous neural oscillations

7.1 Determinism, Compatibilism, and Free Will

It seems to me that I have free will. I have the experience of making choices and enacting those choices in the world. My choices seem to cause events that would not have occurred had I chosen differently. Once I have chosen, I have the distinct

K. B. Laskey (✉)
Systems Engineering and Operations Research Department, George Mason University, Fairfax, VA, USA
e-mail: klaskey@gmu.edu

J. Acacio de Barros, C. Montemayor (eds.), *Quanta and Mind*, Synthese Library 414, https://doi.org/10.1007/978-3-030-21908-6_7

sense that I could have chosen otherwise, and if I had, different outcomes would have occurred. Much of our legal and social structure is rooted in the presumption that we have free will. Our criminal justice and public policy systems treat people as having agency over their lives and responsibility for their choices. Our personal and professional conduct takes for granted that those around us have free will.

But many philosophers and scientists argue that free will is an illusion. The prevailing assumption of Western science is that the behavior of our bodies can be understood entirely in terms of physical law. In this view, thoughts and intentions should be understood as secondary manifestations of the brain's physical state. The assumption of mechanistic materialism typically goes unquestioned in debates between incompatibilists (Harris 2012; Wagner 2003; Smilansky 2000) who think free will is an illusion and compatibilists (e.g., Dennett 1996) who think free will can be reconciled with determinism. Libet's (Libet et al. 1979, 1983) seminal experiments have been taken as confirmation of the illusory nature of free will. In experiments that have been confirmed by a host of follow-on studies, Libet demonstrated that brain activity builds up prior to conscious awareness of undertaking a voluntary behavior. That is, our brains prepare to execute a voluntary action well before we are aware of having chosen it. Psychologists have uncovered numerous ways in which our unconscious minds influence our behavior (e.g., (Bargh 2014). Many have taken these findings to mean that mechanistic materialism is the only scientifically defensible metaphysical stance.

On the other hand, there are benefits to believing in free will. Pearl and Mackenzie argue that the language of free will provides parsimonious encoding of complex causal relationships (Pearl and Mackenzie 2018) that give social agents evolutionary advantage by facilitating effective communication, accelerating learning, and improving collective problem-solving. In addition, there is evidence that belief in free will helps to foster socially adaptive behavior. Belief in free will has been found to correlate with pro-social and ethical attitudes and behaviors (Baumeister and Brewer 2012; Martin et al. 2017). Experiments have shown that undermining subjects' belief in free will tends to increase anti-social and unethical behavior (Baumeister and Brewer 2012). The association between belief in free will and moral judgment is robust across cultures and societies (Martin et al. 2017). Thus, it has been argued, evolutionary pressure favors societies in which individuals believe they have free will. People in such societies communicate better in social situations, learn more rapidly, and engage in pro-social behaviors that foster survival of the group. These findings have led to concern about the effects on society if scientists' skepticism about free will becomes widespread among the general public. "We cannot afford for people to internalize the truth," says the philosopher Smilansky as quoted in (Cave 2016).

Others are less pessimistic about the effects of undermining popular belief in free will. A recent series of studies failed to find a statistically significant effect of general abstract belief in free will on judgments of moral responsibility, but found a strong effect of perceived choice capacity (Monroe et al. 2017). Studies of the lay conception of free will suggest that people focus less on metaphysical notions of indeterminism than on the ability to make reasoned choices without undue influence from external constraints (Monroe and Malle 2010; Stillman et al. 2011). Lay intuitions about free will appear to be multi-faceted, manifesting as compatibilist in some contexts and incompatibilist in others (Nichols and Knobe 2007). These findings suggest the need for further investigation of lay concepts of free will and how belief in free will affects judgments of moral responsibility.

Arguments over compatibilism and the social utility of belief in free will often take as a given that brain processes are primary, mental processes are derivative, and our thoughts have no direct causal influence on the behavior of our bodies. This view is buttressed by the finding that brain activity builds up prior to awareness of a conscious decision, and by research on the strong influence of often unacknowledged unconscious motivations on behavior. Nevertheless, interpreting these findings as definitive proof of mechanistic materialism is premature (e.g., Mele 2009). Libet himself suggested that the intention to perform an action may begin unconsciously, accompanied by a rise in readiness potential, but that conscious choice might still operate to veto the act prior to its occurrence. In this account, consciousness serves to "select and control volitional outcome" that is initiated unconsciously (Libet 1985). According to Lavazza and De Caro (2010):

> ... the application of [experimental neuroscience] findings to the specific but crucial issue of human agency can be considered a "pre-paradigmatic science" (in Thomas Kuhn's sense). This implies that the situation is, at the same time, intellectually stimulating and methodologically confused. More specifically—because of the lack of a solid, unitary and coherent methodological framework as to how to connect neurophysiology and agency—it frequently happens that tentative approaches, bold but very preliminary claims and even clearly flawed interpretations of experimental data are taken for granted.

Based on a review of experimental research and critiques of experimental conclusions, Lavazza and De Caro conclude that many claims of neural determinism are overstated, and that these overstated claims can have negative social consequences. Lavazza (2016) suggests that, given the many theoretical uncertainties and problematic interpretations of experimental findings, a more prudent course is to begin with an operational definition of free will (Lavazza and Inglese 2015) and seek to connect higher-level cognitive constructs to neural processes. Such a research program can be undertaken independently of one's metaphysical stance about whether the brain is actually deterministic and if so, whether determinism can be reconciled with free will.

The psychologist William James (1890) argued against the mechanistic materialist view. His *reductio ad absurdum* lampooned the notion that if we only had an exact description of the brain and environment of Shakespeare, we would be able to show exactly why his hand happened to trace the exact marks we call the manuscript of *Hamlet*. He followed with an evolutionary argument:

> [Consciousness] seems an organ, superadded to the other organs which maintain the animal in the struggle for existence; and the presumption of course is that is helps him in some way in the struggle... But it cannot help him without being in some way efficacious and influencing the course of his bodily history. (James 1890, p. 136)

Walter (2009) defines free will in terms of three properties: alternativism, intelligibility, and origination. Alternativism means the existence of at least two genuinely possible options. Walter means this in the counterfactual sense: having chosen to take a given action, it would have been possible in this instance to have chosen otherwise. Thus, alternativism requires a non-deterministic world. He acknowledges that there can be no definitive proof or refutation of alternativism from empirical observation, because each specific situation occurs only once and can never be duplicated exactly. Nevertheless, he argues, it is meaningful to ask whether, counterfactually, one could have chosen otherwise. Intelligibility means that our choices are made for reasons. That is, rather than choosing randomly or reflexively, we make choices according to our values and for rational reasons. Our minds form representations of the alternatives available to us, predict the consequences of each, evaluate which best serves our values, and choose accordingly. Origination means that we possess agency. As agents, we are the ones who initiate the causal chains that culminate in behaviors we have chosen.

Under the strongest possible interpretation of the terms, Walter argues, alternativism and intelligibility are in conflict with each other. Regardless of whether determinism is true or false, if choices are always made for intelligible reasons, then the choice would have to be the same in any two identical situations, because the same reasons would be operative. Thus, at least one of these properties must be weakened in any non-illusory account of free will. The traditional way to do this is to weaken intelligibility by a conditional argument: I chose this option for given reasons, but I could have chosen otherwise. Walter dismisses this "diluted version" as "not convincing." But is the conditional argument not at the heart of our intuitive conception of free will? We conceive of several options, form representations in our mind of the likely future if each of these options were to occur, consider which of these futures we prefer, and choose the one that seems best. Having chosen, we feel that it would have been possible for us to have chosen otherwise. To say that we are not free because we choose according to our values seems to negate the intuitive meaning of value-driven decision.

Pearl's (2009) *do* calculus provides a useful language for describing the link between our values, our choices, and the consequences of our choices. The *do* calculus is a formal framework for defining and analyzing causal models. A causal model covers both the normal unfolding of events and the effects of "local surgery" operations in which some effects are counterfactually altered. For example, if we learn that the car will not start, we might assign a high probability to a dead battery. That is, the normal unfolding model assigns a high conditional probability to a dead battery given that the car does not start: $\Pr(B = \mathit{dead} | \, C = \mathit{nostart}) = \mathit{HIGH}$. On the other hand, if we intervene to prevent the car from starting without affecting the battery's charge level, we would expect a fully charged battery despite the car's not starting. That is, $\Pr(B = \mathit{dead} | \, \mathit{do}(C = \mathit{nostart})) = \mathit{LOW}$. By contrast, because the battery charge is a cause and not an effect of the car's starting, a causal model would specify both $\Pr(C = \mathit{nostart} \mid B = \mathit{dead}) = \mathit{HIGH}$ and $\Pr(C = \mathit{nostart} \mid \mathit{do}(B = \mathit{dead})) = \mathit{HIGH}$.

The *do* calculus provides a way to formalize Walter's three criteria for free will. Specifically, suppose an agent is given a choice between two actions, $A = a_1$ and $A = a_2$. Alternativism means that both $A = a_1$ and $A = a_2$ are genuine possibilities in this situation. Although Walter gives no formal definition of origination, I propose a formalization in terms of Pearl's *do* calculus. Specifically, origination means that an agent can bring about an intervention $\mathit{do}_{\mathit{Agent}}(A = a)$ to bring about either of the allowable actions. The subscript indicates that it is the agent who can make the action $A = a$ happen. Intelligibility means that the agent's actual choice is made non-arbitrarily. That is, suppose $A = a_1$ will lead to consequence $C = c_1$ and $A = a_2$ will lead to consequence $C = c_2$, and suppose the agent values c_1 more highly than c_2. Intelligibility then means that in the normal course of events, the action $A = a_1$ would occur, because it results in a better consequence according to the agent's value system. However, origination means that both $\mathit{do}(A = a_1)$ and $\mathit{do}(A = a_2)$ are genuine possibilities.

Is this causal model physically possible? Many have asserted the contrary. Clearly, if nature is deterministic, then the intervention $\mathit{do}(A = a_2)$ would not be physically possible without changing the initial conditions. In a deterministic world, our values are encoded in our brain processes, and our bodies automatically enact the more highly valued option $A = a_1$, with no counterfactual alternative being possible in this specific situation. The advent of quantum theory raises the possibility of physical indeterminism. Although there is as yet no consensus on the interpretation of quantum theory, and some popular interpretations are deterministic, there are interpretations in which nature is considered genuinely indeterministic. However, the common conception is that this indeterminism comes in the form of randomness. Because randomness violates intelligibility, quantum randomness cannot rescue free will.

But randomness is not the only way in which indeterminacy enters quantum theory. Another source of indeterminism is the choice of what to measure. This choice is ascribed by the founders of quantum theory to the free will of the scientist. Informal accounts in standard textbooks also use the language of free will in describing the process of observation. Thus, the behavior of a quantum system depends in macroscopically distinguishable ways on an aspect of the world about which the orthodox theory has nothing to say, but that informal accounts ascribe to free will. The ensuing section describes how this choice of what to observe in quantum theory provides an opening for an account of free will that satisfies all three properties of alternativism, intelligibility, and origination. Whether this account is an accurate model of how the world works is as yet an open scientific question. Nevertheless, it is a possibility worthy of the theoretical and empirical work needed to evaluate its merits.

7.2 A Quantum Account of Efficacious Free Choice

Although the predictions of quantum theory have been confirmed to exceptional accuracy, how the theory should be interpreted remains a matter of vigorous debate. Sidestepping this debate, we focus on one specific interpretation that provides the basis for a physically grounded theory of efficacious free choice. Specifically, Stapp (2011, 2017) argues that realistically interpreting von Neumann's (1955) mathematical formalization of quantum theory provides a way for conscious mental effort to become efficacious, while remaining consistent with the empirically well-established rules of quantum theory.

To show how Stapp's interpretation can form the basis of a theory of agency, I begin with an overview of von Neumann's formulation. According to von Neumann, there are two distinct ways in which the state of a quantum system can change over time. The first is deterministic evolution via the Schrödinger equation, which occurs when a quantum system is isolated from its environment. The second rule is a discontinuous change called state reduction, in which the system changes instantaneously to one of a set of possible states. If the state just prior to a reduction and the type of reduction are given, quantum theory specifies probabilities for each of the possible results. The probabilistic predictions of quantum theory have been confirmed to extremely high accuracy. However, quantum theory provides no theory or rules governing the timing or type of reductions.

Therefore, quantum theory is dynamically incomplete. It specifies the behavior of a quantum system conditional on the timing and the type of reductions, but does not specify which reductions are applied at what time. Empirically, reductions are associated with measurements made by scientists, in which a quantum system interacts with a macroscopic measurement device to produce an observable outcome. The lack of a theory for which measurement is performed at what time is known as the *measurement problem*.

The founders of quantum theory stressed that the choice of measurement should be attributed to the free choice of the scientist. Similarly, the informal language describing measurements in quantum theory textbooks associates measurements with choices made by experimenters. But the lack of a fundamental theory of measurement is disconcerting. This insertion of the observer in a fundamental way into the formulation of quantum theory makes it difficult for physicists to stay on their preferred side of the boundary between the physical and the mental (Rosenblum and Kuttner 2011), and has engendered attempts, none fully successful to date, to formulate quantum theory in an observer-independent way.

The laboratory in which a measurement occurs and the body of the experimenter are physical systems, and as such should be governed by quantum theory. To investigate this idea, von Neumann sought to examine the measurement problem within a larger system consisting of the measured and measuring systems. Consider a total system T consisting of a microscopic quantum system Q, a macroscopic measuring device M and an observing scientist O. The usual treatment places a boundary between Q and the entire macroscopic world, i.e. $T = [Q \| M + O]$. On the measured side of the boundary is Q, described in the language of quantum theory, while on the measuring side is $M + O$, described in classical language. Alternatively, suggested von Neumann, we could move the boundary so that M is part of the measured system, obtaining $T = [Q + M \| O]$. We say that O examines the readout of M, which is coupled to the corresponding state of Q. Next, we could move the boundary further to the observer's retina R on which photons bouncing off the measurement device impinge. The observer O is decomposed into $R + O'$, and we now have $T = [Q + M + R \| O^{'}]$. The device M and the retina R are coupled to Q, and O' observes the state of R corresponding to the post-reduction state of Q. We can move further inward to the optic nerve N and to the brain cells B receiving signals from the optic nerve, yielding $T = [Q + M + R + N \| O^{''}]$ and $T = [Q + M + R + N + B \| O^{'''}]$. In each case, the measured system, described in classical language, remains outside of quantum theory. "But in any case," says von Neumann, "no matter how far we calculate ... at some time, we must say: and this is perceived by the observer." The boundary may be arbitrary to some degree, but the world must always be divided into observer and observed, "if the method is not to proceed vacuously, i.e., if a comparison to experiment is to be possible." (von Neumann 1955, p. 420). As long as the measured system proceeds without interacting with the measuring system, the measured system evolves according to the deterministic Schrödinger equation. However, when the system interacts with an observer, a reduction occurs.

The final placement $T = [Q + M + R + N + B \| O^{'''}]$ of the boundary has moved the entire physical world to the side of the measured system, leaving only $O^{'''}$, what von Neumann calls the observer's "abstract ego," on the observing side. The observer's subjective perception of the measurement outcome, says von Neumann, is

> ... a new entity relative to the physical environment and is not reducible to the latter. Indeed, subjective perception leads us into the intellectual inner life of the individual, which is extra-observational by its very nature. (von Neumann 1955, p. 418)

This new entity, which von Neumann identifies with the "inner life of the individual," is as much a part of the world, and as much a valid subject of scientific investigation, as the individual's body and surrounding environment. In fact, our inner life is arguably more basic than our physical bodies. We know incontrovertibly that we have an inner life, but we know about the external world only through our perception of it:

> ... Indeed, experience only makes statements of this type: an observer has made a certain (subjective) observation; and never any like this: a physical quantity has a certain value. (von Neumann 1955, p. 420)

Stapp suggests that we attribute to von Neumann's new entity the ability to effect a reduction to some part of its own physical state. In other words, Stapp hypothesizes that the universe contains a kind of entity that possesses agency, and that agency operates through the initiation of reductions. These new entities, which we might call *reducing agents* (Laskey 2018a, b), initiate reductions to some part of their physical state, and experience the result of the reductions they initiate. Stapp hypothesizes that humans are one kind of reducing agent, but says little about what other kinds of reducing agents might exist in the natural world.

The brain doing the perceiving, says von Neumann, is a physical system, and subjective perception must therefore correspond to a characteristic physical state of the brain:

> … it is a fundamental requirement of the scientific viewpoint – the so-called principle of psycho-physical parallelism – that it must be possible to describe the extra-physical process of the subjective perception as if it were in reality in the physical world – i.e., to assign to its parts equivalent physical processes in the objective environment… (von Neumann 1955, p. 419)

That is, our brains form representations of our past subjective experiences and allow us to make predictions about future experiences. The choices made by a reducing agent are informed by the representations its brain forms of the world around it, which in turn are influenced by sensory inputs and memories of past experiences.

Reducing agents possess all three of Walter's properties of free will. Alternativism is satisfied because there are multiple options available for the timing and type of reductions. Intelligibility is satisfied because reducing agents form representations, make predictions, and select reductions as they deem best. Origination is satisfied because the reducing agent theory ascribes the choice of reductions to the reducing agent.

The hypothesis that the universe contains agents who make free choices by initiating quantum state reductions raises a host of questions. Are humans the only kind of reducing agents? If not, what other kinds of reducing agents are there? Are all quantum state reductions initiated by reducing agents? If not, what distinguishes reductions made by reducing agents from other kinds of reductions? Are all reductions associated with conscious experience? What about the time prior to the evolution of life? Were there reductions then? Were they associated with some variety of consciousness? Any attempt to answer these questions would at present amount to pure speculation. In this paper, I address only the hypotheses that humans are a kind of reducing agent, that humans have free will by virtue of the ability to initiate reductions to relevant parts of our brains, and that consciousness serves to guide our free choices to bring about results we desire. These hypotheses are at present speculative, but as shown in the next section, they are consistent with our current understanding of physics and neuroscience. It cannot yet be known whether the reducing agent hypothesis will ultimately be confirmed by empirical evidence, but it merits scientific investigation.

7.3 Reducing Agents and Neuroscience

Many have argued that quantum theory is not needed for understanding the brain and cognition. Neuroscience research has already helped us to understand many aspects of motor control of volitional action. A network of neurons in our brains processes information via spreading activation. Spreading activation models based on classical physics are studied in neuroscience, robotic control, and machine learning. Most neuroscientists believe that models based on classical physics are adequate for modeling all brain processes, and there is no need to bring in quantum theory.

If the reducing agent hypothesis accounts for human free will, it must be possible for human reducing agents to enact choices in a manner consistent with neurobiology and physiology. Stapp agrees that spreading activation models based on classical physics are adequate for modeling many of the automatic processes associated with cognition, decision-making, and motor control, but that more is needed to model volition. When we prepare for a voluntary action, Stapp suggests, the brain initiates a pattern of neural activity he calls a *template for action* (Stapp 2017). Executing a template for action sends out a sequence of neural impulses instructing the muscles to contract in certain ways that result in characteristic bodily motions. For example, if you are hot and thirsty, your brain might construct a template for action that would cause your arm to reach out to grasp a glass of cold water and bring it to your mouth. Templates for action are fine-tuned with practice to the point where they can be executed smoothly with little conscious attention. Spreading activation models such as those used in neuroscience and robotics can serve as adequate models for the automatic processes involved in retrieving and initiating action templates. Where quantum theory enters, according to Stapp, is the volitional process by which our conscious mind attends to an action template to guide its execution. William James (2001) said, "The essential achievement of the will... is to attend to a difficult object and hold it fast before the mind... Effort of attention is thus the essential phenomenon of will." It is this "effort of attention" where quantum theory plays a role.

Stapp identifies Jamesian "effort of attention" with an essentially quantum phenomenon called the quantum Zeno effect (Misra and Sudarshan 1977), whereby performing very rapid measurements on a quantum system can slow its evolution, in effect "freezing" it in place. The quantum Zeno effect has been confirmed experimentally (Patil et al. 2015). A rapid sequence of reductions can also be used to drive a quantum system along a desired path, in a process that has been dubbed the inverse quantum Zeno effect (Altenmüller and Schenzle 1993). The effect of rapid reduction is fundamentally quantum but does not require the system to be isolated from its environment, and can therefore plausibly operate in brains.

To summarize, Stapp argues that information processing in the brain is well approximated for the most part by a classical spreading activation model. The role of consciousness is to control the rate of reductions, which is the physical manifestation of Jamesian "effort of attention." Exerting effort of attention via the quantum Zeno effect is how free will is hypothesized to operate in humans.

This explanation of how volition acts is consistent with Libet's suggestion that the conscious mind acts to affirm or veto automatically generated behaviors. Because templates for action are constructed automatically, it is natural that brain activity would build up prior to our becoming consciously aware of an action. Consider the above example of reaching for a glass of water. The process is largely unconscious, with conscious attention serving to provide sensory feedback and fine motor guidance. But now consider the situation in which the glass contains a sugary soft drink, and you have resolved to cut down your sugar intake. In this case, you might find yourself beginning to reach for the glass without thinking, and then exerting effort of attention to interrupt this automatic action. At this point, you might actively refocus your attention on a template for turning on the faucet to satisfy your thirst with a glass of water. In this way, you are using Jamesian effort of attention to make the difficult decision to forego a tempting soft drink in favor of a healthier thirst-quenching alternative.

How biologically plausible is this model of conscious volitional action? Where, specifically, in the brain might these quantum effects occur? Several authors have suggested that the distributed, nonlocal nature of the brain's electromagnetic field makes it an attractive candidate as the locus for consciousness (Pockett 2002, McFadden 2013). McFadden (2013) posits that conscious experience involves a feedback loop in which neuronal firing generates an endogenous electromagnetic field in the brain, which in turn influences the rate and synchronicity of neural firing. He points for support to empirical research by Fröhlich and McCormick on the feedback between synchronous firing and the brain's electric field (Fröhlich and McCormick 2010). McFadden's suggestion may be a good model for the automatic aspects of volitional behavior. If so, how might Jamesian effort of attention come into play?

Stapp and others (Stapp 2017; Schwartz et al. 2005; McFadden 2000) have suggested ion channels at nerve terminals as a possible locus of quantum uncertainty in the brain. Schwartz et al. (2005) argue that calcium ion channels are small enough that the uncertainty principle of quantum mechanics comes into play. Release of ions through ion channels at nerve terminals influences the timing of neuron firing. Quantum uncertainty in the ion channels might be important to the precise timing of neuron firing. If so, the quantum Zeno effect might be employed to entrain or disrupt synchronicity of neuronal firing, thereby affecting the strength of the brain's electromagnetic field.

Absent empirical confirmation, this hypothesis must of course be treated as provisional. Nevertheless, it provides an account of free will that satisfies Wagner's three defining properties and is consistent with the laws of physics. At least on preliminary analysis, its account of how efficacious conscious choice is operationalized in biological brains appears to be consistent with neuroscience. Thus, this theory is an existence proof that present-day science cannot rule out the existence of non-illusory free choice. Given the importance of the issue and the concerns expressed by many about the social consequences of widespread disbelief in free will, it is important to acknowledge this existence proof. Research is needed to work out the implications of the theory and to devise empirical tests.

Laskey (2018a, b) suggests a path to empirical evaluation of this theory of efficacious conscious choice. A simple exactly solvable model by Stapp (2017) affirms that the quantum Zeno effect can operate on temporal, spatial and energy scales consistent with neuroscience. Building on this idea, quantum effects could be added to a model if the sort commonly used in computational neuroscience. For example, a model such as the one reported in (Fröhlich and McCormick 2010) could be extended by incorporating quantum uncertainty in ion channels. Because the brain can be modeled to close approximation as a near-classical stochastic process, the computational complexity of such a simulation should be of the same order as a classical neural network simulation. Computational experiments could be performed on the model to investigate whether, in a model with biologically realistic parameter settings, adjusting the rate of reductions can give rise to macroscopic differences in synchronicity. If successful, such simulations could be interleaved with laboratory experiments to investigate whether the model's predictions are borne out in the laboratory.

7.4 Conclusion

The literature on free will rarely questions the underlying metaphysical assumption of mechanistic materialism, regarding it as settled science. Walter (2009) characterizes free will in terms of the properties of alternativism, intelligibility, and origination. He argues that under their strongest interpretation, the three properties are not consistent with each other. His strong interpretation does not conform to the intuitive notion of free will as the *ability* to choose according to rational analysis, not the *necessity* that we do so. Stapp's realistic interpretation of quantum theory satisfies a less stringent interpretation of Walter's properties and provides an opening for free will to operate in a manner fully consistent with the laws of physics. Stapp postulates the existence of a new kind of entity in the natural world. These new entities, dubbed *reducing agents* by Laskey, are able to choose, within as yet to be determined physical limits, when to initiate quantum state reductions to some part of their physical state. Reducing agents are identified with the "new kind of entity" von Neumann associated with the "intellectual inner life of the individual." In Stapp's model of efficacious conscious choice, the brain constructs a template for action by a process that can be well approximated by a classical stochastic neural network model. Quantum theory enters through the application of Jamesian attention density, which employs the quantum Zeno effect to hold a desired template for action in place longer than it would persist otherwise.

The reducing agent theory of efficacious conscious choice is consistent with the accepted laws of physics, but must be regarded as provisional until empirical evidence either confirms or refutes it. Whether this theory of agency proves correct or turns out to be a scientific dead end, its profound implications argue for taking it seriously enough to devise and conduct tests of its plausibility. A route to evaluating the theory is to implement a simulation of a neural network model in which the quantum Zeno effect acts to reinforce or disrupt the execution of neural action templates, and to perform computational experiments to evaluate whether different attention density settings lead to macroscopically distinguishable differences in behavior with neurologically plausible parameter settings. If successful, the theory could be refined through a process of comparing predictions to laboratory experiments and adjusting the model to make it more realistic.

Acknowledgements Acknowledgement is due to Henry Stapp for extensive discussions, explanations, and feedback as the ideas expressed in this paper evolved. Appreciation is extended to participants in the April 2018 International Conference on Quanta and the Mind for stimulating discussions.

References

Altenmüller, T. P., & Schenzle, A. (1993, July). Dynamics by measurement: Aharonov's inverse quantum Zeno effect. *Physical Review. A, Atomic, Molecular, and Optical Physics, 48*(1), 70–79.

Bargh, J., A. (2014, January). How unconscious thought and perception affect our every waking moment. *Scientific American*, 32–39.

Baumeister, R. F., & Brewer, L. E. (2012). Believing versus disbelieving in free will: Correlates and consequences. *Social and Personality Psychology Compass, 6*(10), 736–745.

Cave, S. (2016, June). There's no such thing as free will. *The Atlantic* [Internet]. [cited 2018 Aug 23]. Available from: https://www.theatlantic.com/magazine/archive/2016/06/theres-no-such-thing-as-free-will/480750/

Dennett, D. C. (1996, January 1). Facing backwards on the problem of consciousness. *Journal of Consciousness Studies, 3*(1), 4–6.

Fröhlich, F., & McCormick, D. A. (2010, July 15). Endogenous electric fields may guide neocortical network activity. *Neuron, 67*(1), 129–143.

Harris, S. (2012). *Free will* (1st ed.). New York: Free Press.

James, W. (1890). *Principles of psychology*. New York: Dover.

James, W. (2001). *Psychology: The briefer course. Later edition* (p. 368). Mineola: Dover Publications.

Laskey, K. B. (2018a). Acting in the world: A physical model of free choice. *Journal of Cognitive Science, 19*(2), 125–163.

Laskey, K. B. (2018b, October 6). A theory of physically embodied and causally effective agency. *Information, 9*(10), 249.

Lavazza, A. (2016, June 1). Free will and neuroscience: From explaining freedom away to new ways of operationalizing and measuring it. *Frontiers in Human Neuroscience*, 10. [Internet. Cited 2018 August 26]. Available from: https://www.ncbi.nlm.nih.gov/pmc/articles/PMC4887467/.

Lavazza, A., & De Caro, M. (2010, April 1). Not so fast. On some bold neuroscientific claims concerning human agency. *Neuroethics, 3*(1), 23–41.

Lavazza, A., & Inglese, S. (2015, April 17). Operationalizing and measuring (a kind of) free will (and responsibility). Towards a new framework for psychology, ethics, and law. *Rivista internazionale di Filosofia e Psicologia, 6*(1), 37–55.

Libet, B. (1985). Unconscious cerebral initiative and the role of conscious will in voluntary action. *The Behavioral and Brain Sciences, 8*(4), 529–566.

Libet, B., Wright, E. W. J., Feinstein, B., & Pearl, D. K. (1979). Subjective referral of the timing for a conscious sensory experience. *Brain, 102*, 193–224.

Libet, B., Gleason, C. A., Wright, E. W., & Pearl, D. K. (1983, September). Time of conscious intention to act in relation to onset of cerebral activity (readiness-potential). The unconscious initiation of a freely voluntary act. *Brain: A Journal of Neurology, 106*(Pt 3), 623–642.

Martin, N. D., Rigoni, D., & Vohs, K. D. (2017, July 11). Free will beliefs predict attitudes toward unethical behavior and criminal punishment. *Proceedings of the National Academy of Sciences, 114*(28), 7325–7330.

McFadden, J. (2000). *Quantum evolution*. New York: Norton.

McFadden, J. (2013). The CEMI field theory closing the loop. *Journal of Consciousness Studies, 20*(1–2), 1–2.

Mele, A. R. (2009). *Effective intentions: The power of conscious will* (p. 208). Oxford/New York: Oxford University Press.

Misra, B., & Sudarshan, E. C. G. (1977, April 1). The Zeno's paradox in quantum theory. *Journal of Mathematical Physics, 18*, 756–763.

Monroe, A. E., & Malle, B. F. (2010, June 1). From uncaused will to conscious choice: The need to study, not speculate about people's folk concept of free will. *Review of Philosophy and Psychology, 1*(2), 211–224.

Monroe, A. E., Brady, G. L., & Malle, B. F. (2017, March 1). This isn't the free will worth looking for: General free will beliefs do not influence moral judgments, agent-specific choice ascriptions do. *Social Psychological and Personality Science, 8*(2), 191–199.

Nichols, S., & Knobe, J. (2007). Moral responsibility and determinism: The cognitive science of folk intuitions. *Noûs, 41*(4), 663–685.

Patil, Y. S., Chakram, S., & Vengalattore, M. (2015, October 2). Measurement-induced localization of an ultracold lattice gas. *Physical Review Letters, 115*(14), 140402.

Pearl, J. (2009). *Causality: Models, reasoning, and inference* (2nd ed.). Cambridge/New York: Cambridge University Press.

Pearl, J., & Mackenzie, D. (2018). *The book of why: The new science of cause and effect* (1st ed., p. 432). New York: Basic Books.

Pockett, S. (2002). On subjective back-referral and how long it takes to become conscious of a stimulus: A reinterpretation of Libet's data. *Consciousness and Cognition, 11*(2), 141–161.

Rosenblum, B., & Kuttner, F. (2011). *Quantum enigma: Physics encounters consciousness* (2nd ed., p. 304). Oxford/New York: Oxford University Press.

Schwartz, J. M., Stapp, H. P., & Beauregard, M. (2005). Quantum physics in neuroscience and psychology: A new model with respect to mind/brain interaction. *Philosophical Transactions of the Royal Society B, 360*(1458), 1309–1327.

Smilansky, S. (2000). *Free will and illusion* (1st ed., p. 344). Oxford/New York: Oxford University Press.

Stapp, H. P. (2011). *Mindful universe: Quantum mechanics and the participating observer* (2nd ed., p. 230). Berlin/New York: Springer.

Stapp, H. P. (2017). *Quantum theory and free will: How mental intentions translate into bodily actions* (1st ed., p. 142). New York: Springer.

Stillman, T. F., Baumeister, R. F., & Mele, A. R. (2011). Free will in everyday life: Autobiographical accounts of free and unfree actions. *Philosophical Psychology, 24*(3), 381–394.

von Neumann, J. (1955). *Mathematical foundations of quantum mechanics*. Princeton: Princeton University Press.

Wagner, D. M. (2003). *The illusion of conscious will*. Cambridge, MA: Bradford Books.

Walter, H. (2009). *Neurophilosophy of free will: From libertarian illusions to a concept of natural autonomy* (p. 420). MIT Press.

Chapter 8
Bohmian Philosophy of Mind?

Peter J. Lewis

Abstract It has been suggested by some proponents of Bohm's theory that it requires a special account of mental awareness–that the Bohmian solution to the measurement problem rests on direct awareness of the particle configuration in one's own brain. This suggestion leads to two criticisms of Bohm's theory: first, that direct awareness of particle configuration is a highly implausible account of the mental; and second, that such direct awareness leads to violations of the quantum no-signalling theorem. I argue that Bohm's theory requires no special account of mental awareness, and hence that neither of these problems arises.

8.1 Introduction

Bohm's theory is in many ways an attractive solution to the measurement problem in quantum mechanics. It provides an intuitive explanation for the distinctive quantum phenomena of interference and entanglement without the need for any problematic "collapse" of the wave function. But it faces several serious difficulties. First, the dynamical law via which the wave function "pushes around" the Bohmian particles is explicitly non-local, against the spirit of special relativity (Bell 1987, 115). Second, the Bohmian particles can be seen as *redundant* in the context of an Everettian solution to the measurement problem (Brown and Wallace 2005). And third, the Bohmian solution to the measurement problem apparently depends on an implausible and problematic account of mental awareness (Stone 1994; Brown and Wallace 2005).

I do not wish to minimize the significance of the first two difficulties; they are serious threats to the tenability of Bohm's theory. But the third difficulty, I think, rests on a confusion concerning the way in which Bohmian particles encode the outcomes of measurements. In particular, my concern here is to respond to the

P. J. Lewis (✉)
Dartmouth College, Hanover, NH, USA
e-mail: Peter.J.Lewis@dartmouth.edu

J. Acacio de Barros, C. Montemayor (eds.), *Quanta and Mind*, Synthese Library 414, https://doi.org/10.1007/978-3-030-21908-6_8

accusations of Stone (1994) and Brown and Wallace (2005) that Bohm's theory requires a mysterious kind of direct awareness of the positions of the Bohmian particles in our brains, and also to the claim of Brown and Wallace (2005) that such direct awareness threatens the quantum no-signaling theorem.

8.2 The Case Against Bohm

The background to the critique I wish to discuss is an influential discussion of Bohm's theory by Dürr, Goldstein and Zanghì (1992). Dürr et al. derive a result they call "absolute uncertainty", which says that "the quantum equilibrium hypothesis $\rho = |\psi|^2$ conveys the most detailed knowledge possible concerning the present configuration of a subsystem (of which the "observer" or "knower" is not a part)" (1992, 882). The Bohmian particle configuration is always precisely defined, so it might seem that it should be possible to find out what that particle configuration is. Dürr et al.'s "absolute uncertainty" result apparently says that we *can't* find out: "no devices whatsoever, based on any present or future technology, will provide us with the corresponding knowledge. *In a Bohmian universe such knowledge is absolutely unattainable!*" (1992, 882). The best we can do is to assign a probability distribution ρ to the possible particle configurations given by the squared wave function amplitude $|\psi|^2$.

Dürr et al. intend this result as a *defense*, not a criticism, of Bohm's theory. Indeed, it is central to the empirical adequacy of Bohm's theory that the probability distribution over particle configurations is $|\psi|^2$, in accordance with the Born rule. Nevertheless, the "absolute unattainability" of knowledge of the particle configuration is the source of a recurring objection to Bohm's theory.

The original source of the objection is Stone (1994). Stone argues that the Bohmian particle configuration contains no information. In a sense, of course, the Bohmian particle configuration certainly *does* contain information, namely the precise values of three coordinates for each particle in the system. But if we take "information" to mean "*accessible* information", then arguably the Dürr et al. result entails that the Bohmian particle configuration contains no information over and above the wave function distribution, since that result seems to imply that we can't *find out* the particle configuration with any greater accuracy than $|\psi|^2$. Hence Stone concludes that "we will never have an empirical way to decide between the many competing stories about the Bohm trajectories" (1994, 264). Since learning that one among the many possible Bohm trajectories is actual is central to the Bohmian solution to the measurement problem, Stone concludes that Bohm's theory fails to solve the measurement problem.

It is worth noting, though, that Dürr et al.'s result includes the caveat that the observer is not part of the system. In a footnote, they expand on this exception: "There is one situation where we may, in fact, know more about configurations than what is conveyed by the quantum equilibrium hypothesis $\rho = |\psi|^2$: when we ourselves are part of the system!" (1992, 903). Stone takes this to mean that "while detailed knowledge of the configurational states of external systems is forever unattainable to us, we can nevertheless have knowledge of (perhaps we should say 'have our knowledge in') the configuration of our own particles" (1994, 264). Here we get the first inkling of the suggestion that Bohm's theory requires a distinctive account of mental awareness—that *direct awareness* of the particle configuration in our own brains can bypass the prohibition on knowing the particle configuration, and hence provide a route via which Bohm's theory can solve the measurement problem.

Stone immediately objects to this suggestion: "What physical model of brain processes can possibly underlie this statement? Suppose I consider a single neuron of my brain as "the system" and the rest of my brain as part of the "environment". Since the "environment" does not contain information about the "system" configuration (beyond what is available from its wave function), whatever knowledge this neuron may have stored up in the configuration of its particles is "absolutely unattainable" to the rest of my brain!" (1994, 264). That is, Dürr et al.'s result entails that any "direct awareness" of the particle configuration by one part of my brain would be completely inaccessible by the rest of my brain. Stone is surely right to think that any such hermetically sealed "awareness" couldn't in principle fulfill the functions of *genuine* awareness in guiding belief and action.

So with or without the proposed exception for direct awareness, Stone concludes that Bohm's theory is incapable of solving the measurement problem. But is this criticism correct? Maudlin takes issue with Stone's attack, arguing that "the obvious answer to his complaint is that no one ever showed that in Bohm's theory particle positions cannot store information about other particle positions, only that *at the beginning of a measurement* the positions of particles in the environment store no more information about the particles in the measured system *than is reflected in the effective wave function* (1995, 481). That is, Maudlin accuses Stone of misinterpreting Dürr et al.'s result, reading a prohibition on finding out the particle position *prior to* a measurement as an *absolute* prohibition.

How, then, does Maudlin think that we can find out the position of a Bohmian particle? Simply by correlating its position to the positions of *other* Bohmian particles: "If we want to know more, we couple the system to a measuring device which correlates the positions of particles in the measured system to those in the measuring system" (1995, 483). And how do we find out the positions of *those* particles? "If we want to know what happened to the measuring device (e.g., which way the pointer went), we look at it, thereby correlating positions of particles in our brains with the pointer position" (1995, 483). Hence we gain information about the Bohmian particle configuration: "If getting the state of our brain correlated with previously unknown external conditions is not getting information about the world, then nothing is" (1995, 483).

Maudlin concludes that "Bohm's theory solves the measurement problem completely and without remainder" (1995, 483). But not everyone is convinced. After all, Maudlin's solution to Stone's worry about the accessibility of particle positions is to insist that we can know them via *other* particle positions; but if particle positions *in general* are inaccessible, this is no help. Perhaps, then, it is the particle positions *in our brains* that are doing the work here in giving us access.

This is how Brown and Wallace interpret Maudlin: "Maudlin seems to be taking it for granted that our conscious perceptions supervene directly and exclusively on the configuration of (some subset) of the corpuscles associated with our brain" (2005, 534). But why think that we are directly aware of the configuration of particles in some part of our brain? After all, this seems to lead us right back Stone's concerns that such "awareness" would be inaccessible to the rest of our brain. Furthermore, it seems to involve us in "the assumption that consciousness is some sort of bare physical property (like, say, charge)," which "makes consciousness completely divorced from any assumptions rooted in the study of the brain" (2005, 536). Finally, "a violation of the no-signaling theorem is possible in principle were we to 'know' the configuration of corpuscles in our brain with a greater level of accuracy than that defined by the wave function" (2005, 535).

In sum, then, the gist of the critique of Bohm's theory is this: Bohmian particle positions, which are key to solving the measurement problem, are in general unknowable. In fact, the *only* way we might know them is via direct awareness in our brains. But this is a heterodox and highly implausible account of the nature of awareness, and what's more, it threatens the no-signaling theorem, which is important for the reconciliation of quantum mechanics and special relativity.

8.3 How to Send a Superluminal Signal

Let us consider the last point more carefully. Why would knowing the Bohmian particle configuration allow one to send a superluminal signal, and why does it matter? Consider two spin-1/2 particles in the entangled state $2^{-1/2}(|\uparrow\rangle_A\,|\downarrow\rangle_B - |\downarrow\rangle_A\,|\uparrow\rangle_B)$. Suppose that Alice takes particle A and Bob takes particle B, and they perform spin measurements on their respective particles at space-like separated locations. As Bell (1987, 14) showed, the results of their measurements will exhibit correlations that can't be explained by positing local, intrinsic properties of the individual particles. It seems that quantum entanglement involves us in some kind of non-locality or non-separability or holism—some kind of "direct link" between the two particles, no matter how far apart they are.

Nevertheless, it can be shown that according to standard quantum mechanics, there is nothing that Alice can do to her particle that could be used to send a signal to Bob. This is important because it suggests the possibility of a peaceful coexistence of quantum mechanics and special relativity: while a "direct link" between space-like separated events may be in tension with the spirit of special relativity, there is arguably no outright violation of special relativity absent a superluminal signal.

Insofar as Bohm's theory is empirically equivalent to standard quantum mechanics, the no-signaling theorem is retained. So as long as Alice can know no more about her particle than is given by the Born rule, then she cannot send a signal to Bob. But suppose that Alice *can* know the location of her particle in its wave packet with greater precision than $|\psi|^2$: then she can send a signal.

To see how this is possible, consider how the state of the system evolves as Alice and Bob make their measurements. The easiest way for each of them to measure the

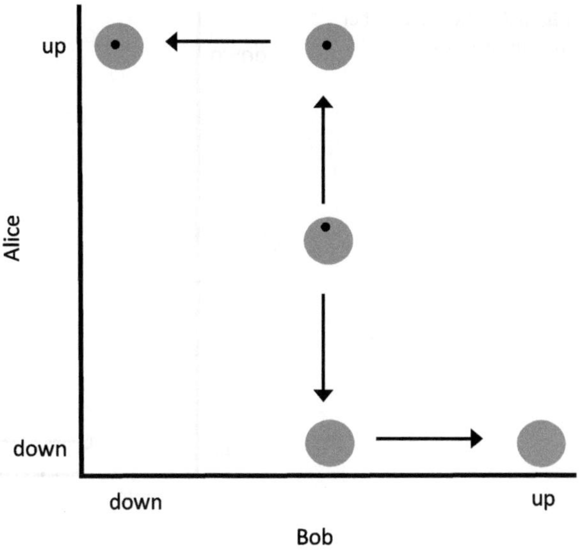

Fig. 8.1 Spin measurements on entangled particles

spin of their particle is to pass it through an inhomogeneous magnetic field oriented along some chosen axis, and then run it into a fluorescent screen that lights up at the point of contact. If Alice and Bob orient their magnets in the same direction—say along the *z*-axis—the measurements can be represented as in Fig. 8.1. For a two-particle system, the wave function inhabits a six-dimensional configuration space, but for picturability, we can focus on the *z*-coordinate of Alice's particle, plotted vertically, and the *z*-coordinate of Bob's particle, plotted horizontally. The circle represents the region of configuration space in which the wave function amplitude is large, and the point represents the positions of the two Bohmian particles.

Consider a frame of reference in which Alice's measurement occurs first. As her wave packet passes through the magnetic field, it splits into two components based on its spin—a spin-up wave packet displaced upwards, and a spin-down wave packet displaced downwards. The Bohmian particle follows one of the components, depending on its initial position: if it is above the midpoint of the initial wave packet in Alice's *z*-coordinate, it moves upwards, and otherwise it moves downwards.[1]

Now Bob passes his wave packet though a magnetic field. Given the entangled nature of the original state, the wave packet that is spin-up for Alice's particle is spin-down for Bob's particle, and vice versa. Hence there is no further splitting of the wave packets in configuration space: the packet that was deflected upwards in Alice's *z*-coordinate is deflected downwards in Bob's *z*-coordinate, and vice versa. The Bohmian particle is carried along with the packet it occupies. Hence if Alice's

[1]For a single particle, this is because otherwise Bohmian trajectories starting below the midpoint and above the midpoint would intersect each other, and intersecting trajectories are prohibited in a deterministic theory. For two particles in a six-dimensional configuration space, there is no danger of the trajectories intersecting, but the additional degrees of freedom corresponding to Bob's particle are irrelevant to the motion of Alice's particle.

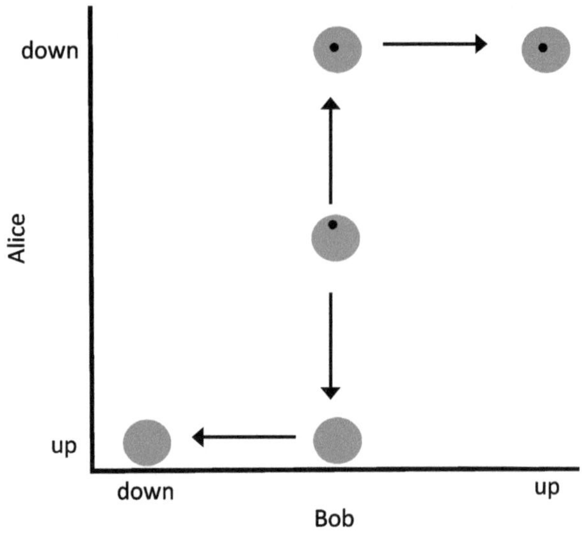

Fig. 8.2 Alice rotates her measuring device

particle is above the midpoint in the initial wave packet, as shown in Fig. 8.1, then Alice gets the result "spin-up" for her measurement and Bob gets the result "spin-down".

Figure 8.2 shows what happens if Alice rotates her measuring device by 180°. Now the spin-up component of Alice's wave packet is displaced downwards, and the spin-down component is displaced upwards. But as before, if the Bohmian particle is above the midpoint of the initial wave packet in Alice's z-coordinate, it moves upwards, and otherwise it moves downwards.[2] When Bob passes his wave packet through the magnetic field, the packet that was deflected upwards in Alice's z-coordinate is deflected downwards in Bob's z-coordinate, and vice versa, and the Bohmian particle goes with it. Hence if Alice's particle is above the midpoint in the initial wave packet, as shown in Fig. 8.2, then Alice gets the result "spin-down" for her measurement and Bob gets the result "spin-up".

Note that for the same initial state (wave packet plus Bohmian particle position), the results of the measurements depend on the orientation of Alice's measuring device. One way up, Alice gets "spin-up" and Bob gets "spin-down". The other way up, Alice gets "spin-down" and Bob gets "spin-up". This is an illustration of the contextuality of spin in Bohm's theory: the result of a spin measurement depends on how that spin is measured. But if Alice can locate her particle with greater accuracy than $|\psi|^2$, it also provides a way for Alice to send a superluminal signal to Bob. All she needs to do is to observe whether her particle is above or below the midpoint of her wave packet. If it is above the midpoint, then to send the signal "spin-up" to Bob she rotates her measuring device, and to send "spin-down" she leaves it as it is. If her particle is below the midpoint, she reverses this strategy.

[2]By the same argument as before.

Perhaps, though, Alice is only directly aware of the positions of particles in her own brain. Even so, if it can be arranged that Alice's particle in the above experiment is embedded in her brain in the relevant way, then she can use her direct awareness of the position of this particle to send a signal to Bob. That is, it looks like Brown and Wallace are correct that direct awareness of the positions of the Bohmian particles threatens the no-signaling theorem, and hence the possibility of peaceful coexistence of quantum mechanics and special relativity.

8.4 Awareness as a Red Herring

Stone (1994, 264) and Brown and Wallace (2005, 534) each complain that making direct awareness *exceptional* in this way—allowing that one can be directly aware of the position of a Bohmian particle even though no other physical process can locate a Bohmian particle with greater accuracy than $|\psi|^2$—threatens standard assumptions about the nature of mind. If minds are physically instantiated, how can they operate in ways that other physical systems cannot? In particular, how can a particle embedded in Alice's *brain* be used to send a superluminal signal, when a similar set-up outside her brain cannot? It all looks decidedly spooky.

However, I think all this talk about what Alice is directly aware of is a distraction. There is a perfectly straightforward sense in which the positions of Bohmian particles encode *accessible information* about the outcomes of measurements, contra Stone. And the ability to access this information wouldn't give a system (or a person) the ability to send a superluminal signal, contra Brown and Wallace.

Let's start with the first point: Bohmian particles encode accessible information. As mentioned previously, there is a trivial sense in which the Bohmian particle configuration contains information that is not contained in the wave function. Consider the initial position of the Bohmian particle in Fig. 8.1. The wave function is entirely symmetric around the midpoint of Alice and Bob's z-coordinates, but the particle configuration breaks the symmetry. Furthermore, the particle configuration is *predictive* of the result of the spin measurement: if Alice's particle is above the midpoint in her z-coordinate, Alice gets spin-up and Bob gets spin-down, whereas if the particle is below the midpoint, Alice gets spin-down and Bob gets spin-up. Finally, at the end of the measurement, the particle configuration is *perfectly indicative* of—one might even say *constitutive of*—the outcome.

Hence there is an obvious sense in which the particle configuration contains information about measurement outcomes, information that is accessible via measurements. Why think otherwise? There are a number of concerns one might have. First, the above story depends on Alice performing her measurement *before* Bob, but given that the measurement locations are space-like separated, there is (according to special relativity) no fact about which measurement is performed first. This is an entirely reasonable criticism of Bohm's theory: the dynamics of the theory are explicitly non-local, and require an absolute standard of simultaneity in order to be well-defined. But given that this prerequisite of the Bohmian dynamics is satisfied, it makes sense to say that Alice's measurement occurs first.

Second, the above story is relative to an orientation of Alice's measuring device: if she rotates her device by 180°, the particle configuration contains *different* information about the measurement outcomes. This is an expression of the well-known *contextuality* of properties other than position in Bohm's theory: spin is not an intrinsic property of a particle, but is defined only relative to a measurement context. Even so, *granted* this contextuality, the particle configuration contains accessible information about the (contextually defined) spin properties of the particles.

Finally, and most importantly for present concerns, it might be objected that the above story begs the question, in that it assumes that the particle configuration at the end of the measurement—the one I said was constitutive of the outcome—is *accessible*. Doesn't the Dürr et al. "absolute uncertainty" result show that knowledge of the particle configuration is "absolutely unattainable"? I could spin out the story further, but the objection would recur. If the particles are detected by running them into a fluorescent screen, then the position of the measured spin-1/2 particle is reflected in the positions of the electrons in the excited atoms at the impact point. If the light from the excited atoms is detected, then the positions of the particles in the display of the photon detector reflect the position of the measured spin-1/2 particle. At each stage, the position (and hence the spin) of our original Bohmian particle becomes correlated with the positions of more and more Bohmian particles in the environment. But if the positions of Bohmian particles are in general inaccessible, how does this help?

The answer, I think, is to appeal to functionalism. The spin of the original particle is correlated with the positions of the particles in the photon detector. The positions of those particles can in turn be used to control further physical systems in any way whatsoever. That is, the Bohmian particles in the photon detector can access the spin of the original particle on any reasonable functional characterization of what it takes to access an aspect of the physical world.

This is essentially Maudlin's (1995, 483) answer. But Maudlin (unintentionally) muddies the water by spinning out the story in terms of a correlation with particles in an observer's *brain*. This in turn leads Brown and Wallace (2005, 534) to conclude that Maudlin is appealing to some special account of direct awareness. But as I hope to have shown here, there is no need to mention either brains or awareness: *any* system can in principle access the spin of the particle and use it to control other systems.

The appeal to functionalism is this context is something of a double-edged sword, however. The main argument of Brown and Wallace (2005) is that the *wave function* can perform all the functions that the Bohmian particle configuration can perform, and hence that the Bohmian particles are *redundant*. The difference, of course, is that the Bohmian particle configuration picks out *one* result of the spin measurement, whereas the wave function is symmetric between *all* possible results. But provided that an Everettian or many-worlds solution to the measurement problem is tenable, the Bohmian particle configuration arguably adds nothing.

As I mentioned earlier, I do not mean to dismiss this redundancy argument. If there is a response, it is that the Everettian solution to the measurement problem might *not* be tenable (Callender 2010). But for present purposes, my point is that, setting aside considerations of redundancy and of non-locality, there is no *additional* problem concerning the accessibility of the Bohmian particles.

What of Dürr et al.'s "absolute uncertainty" result, then? The key here, as Maudlin correctly notes, is that when Dürr et al. say that we can't know the particle configuration with greater precision than $\rho = |\psi|^2$, the ψ in question is the *effective* wave function. The effective wave function is the component of the wave function that is relevant to our concerns, *given* what we know. And in the context of Bohm's theory, the effective wave function is the component of the wave function that is relevant to our concerns, given what we know *of the Bohmian particle configuration*.

Consider again Alice's spin measurement. According to Bohm's theory, the wave function never collapses, so the quantum state of the world will be very complicated indeed. But given the effects of decoherence, that state will naturally decompose into a number of branches, most of which are irrelevant to the behavior of the branch containing the Bohmian particles. If the experiment has been set up correctly, the branch containing the Bohmian particles will take the form of the entangled wave packet we have been studying. This is the effective wave function at the beginning of the measurement. Dürr et al.'s result entails that at the beginning of the measurement, Alice knows no more about the Bohmian particle configuration than that it has a probability distribution given by the absolute square of this effective wave function.

But what about at the end of the measurement? Here again Maudlin inadvertently muddies the waters by stressing that Dürr et al.'s result applies to Alice at the *beginning* of the measurement, perhaps implying to some readers that at the end of the measurement we can know the particle configuration with *more* accuracy than given by the square of the effective wave function. Indeed, he adds that "if we want to know more, we couple the system to a measuring device... If we want to know what happened to the measuring device (e.g., which way the pointer went), we look at it, thereby correlating positions of particles in our brains with the pointer position" (1995, 483). This might inadvertently suggest that particles in our brains have a special role in allowing us to know more than the square of the effective wave function.

Of course, it is entirely correct to say that at the end of the measurement we know more than at the beginning. But the crucial point is that Dürr et al.'s result applies equally at the end of the measurement—it's just that the effective wave function has *changed*. When Alice learns that the result of her measurement is spin-up, she learns that the Bohmian particle is not associated with the spin-down wave packet, and hence that she can ignore it. That is, the effective wave function changes from the entire entangled state to just one term in this state.

Does she know more than is given by the squared wave amplitude of this remaining term in the wave function? Well, maybe—it depends on the accuracy of the measurement via which she locates the particle. Perhaps she performs a very rough position measurement that only distinguishes the spin-up term from the spin-down term, and nothing more; in that case, the post-measurement effective wave function is just the spin-up term. Or perhaps the position measurement is more accurate; in that case the post-measurement effective wave function is more tightly localized. The point is that no measurement is *completely* accurate, and however tightly localized the final effective wave function turns out to be, Alice's knowledge of the position of the Bohmian particle will be distributed according to the square of this effective wave function.

So Dürr et al.'s result applies both before and after a measurement, and in no way precludes finding out about the Bohmian particle configuration. Seen in this way, the result might look trivial: the post-measurement effective wave function reflects what you *know* of the Bohmian particle configuration, so *by definition* you can't know the configuration with greater accuracy! But it is nevertheless an important result: it shows that Bohm's theory is *consistent*. The wave function plays a peculiar dual role in Bohm's theory: a dynamical role in pushing the particles around, *and* an epistemic role in reflecting our knowledge of the particle configuration. It is important that these roles always coincide, and Dürr et al.'s result shows that they do: the relevant part of the wave function, dynamically speaking, is always also the part over which your knowledge of the particle configuration is distributed according to $\rho = |\psi|^2$.

Nevertheless, Dürr et al. let their rhetoric get away with them. Knowledge of the Bohmian particle configuration is not "absolutely unattainable"—in fact, knowledge of the particle configuration is easily attainable by a simple position measurement. Perhaps what they mean is that no measurement is *perfectly* accurate, so one can never know the Bohmian particle configuration with perfect precision. But this is hardly a surprise; it is equally true of the classical particle configuration.

Similarly, Dürr et al. are mistaken in suggesting that there is an exception to their result for knowledge of the particles in your own brain. As Stone correctly points out, in order to count as *knowledge*, such self-awareness needs to be accessible by other parts of your brain, and any physical account of this process will be subject to Dürr et al.'s result. Whatever you find out about the configuration of Bohmian particles in your own brain via such a process, you will never attain *perfect* precision, and your probability distribution over the possible particle configurations will be given by the squared amplitude of the effective wave function—the wave function *given* what you have found out.

The lack of an exception is good news for Bohm's theory, in that it avoids the criticisms of Stone and of Brown and Wallace directed at the exception. Bohm's theory does not need any special account of awareness of your own brain state. Nevertheless, you can find out about the Bohmian particle configuration, both inside and outside your brain, to any practicable degree of accuracy. What, then, of Brown and Wallace's further contention that such knowledge would allow one to send a superluminal signal?

8.5 No Signaling

Consider again what Alice needs to do to send a superluminal signal. She needs to find out the position of the Bohmian particle relative to her wave packet, and set her measuring device accordingly. As detailed above, it is perfectly possible for her to find out whether the Bohmian particle is above or below the midpoint of the wave packet: she simply needs to perform the relevant measurement. The relevant measurement in this case is to pass the wave packet through a magnetic field and then to detect whether the particle moves up or down. If it moves up, she now knows that it was above the midpoint.

But of course by this stage it is *too late* to set her measuring device according to the initial position of the particle: she has already *measured* her particle, and in doing so, has moved it from its initial position. In other words, in order to send a superluminal signal, Alice would have to act on the particle position *before* she has accessed that position. Trivially, Alice can't do that, even though she can perfectly well find out the position of her particle to any practicable degree of accuracy. Her ability to find out the Bohmian particle configuration does not allow her to send a superluminal signal.

8.6 Conclusion

Stone contends that Bohm's theory doesn't solve the measurement problem, because Dürr et al.'s result means that you can never find out the Bohmian particle configuration, except perhaps via some implausible direct awareness of your own brain state. Stone's argument rests on a mistaken reading of Dürr et al.'s result, albeit one that is suggested by some of Dürr et al.'s rhetoric. Maudlin correctly identifies Stone's error, but continues to suggest (perhaps inadvertently) that there is some special role for awareness of your own brain state in finding out the particle configuration. I hope to have shown here that there is no special problem in finding out the Bohmian particle configuration, and that acquiring such information neither conflicts with Dürr et al.'s result, nor requires any special role for direct awareness of your own brain state. Finally, finding out the Bohmian particle configuration does not allow you to send a superluminal signal, as Brown and Wallace contend. In short, Bohm's theory provides a perfectly straightforward solution to the measurement problem, and one that does not require any special account of mental awareness.

References

Bell, J. S. (1987). *Speakable and unspeakable in quantum mechanics*. Cambridge: Cambridge University Press.

Brown, H., & Wallace, D. (2005). Solving the measurement problem: De Broglie–Bohm loses out to Everett. *Foundations of Physics, 35*, 517–540.

Callender, C. (2010). *Discussion: The redundancy argument against Bohm's theory*. Unpublished manuscript.

Dürr, D., Goldstein, S., & Zanghì, N. (1992). Quantum equilibrium and the origin of absolute uncertainty. *Journal of Statistical Physics, 67*, 843–907.

Maudlin, T. (1995). Why Bohm's theory solves the measurement problem. *Philosophy of Science, 62*, 479–483.

Stone, A. (1994). Does Bohm's theory solve the measurement problem? *Philosophy of Science, 61*, 250–266.

Chapter 9
Mind and Matter. Two Entangled Parallel Time-Lines, One Reconstructing the Past in Remembering, the Other Extrapolating into the Future in Predicting

Giuseppe Vitiello

Abstract The emergence of collective behavior of myriads of neurons forming coherent patterns of amplitude modulated oscillations is described in terms of quantum field theory formalism with spontaneous breakdown of symmetry. The dissipative quantum model of brain and its results in agreement with observations are discussed. Mental activity and brain activity cannot be separated. In the many-body model they are dynamically entangled. The act of consciousness might reside in such an entanglement.

Keywords Brain functional activity · Mind · Symmetry breakdown · Coherence · Quantum field theory

Walter Freeman observed in his laboratory "the tendency of vast collections of neurons to shift abruptly and simultaneously from one complex activity pattern to another in response to the smallest of inputs" (Freeman 1991). According to him, this behavior "underlies the ability of the brain to respond flexibly to the outside world and to generate novel activity patterns, including those that are experienced as fresh ideas." (Freeman 1991). He also observed that same stimulus triggering these assemblies of myriads of neurons organized in synchronized amplitude modulated (AM) phase-locked oscillations would not produce, in same conditions, the same neuronal oscillation patterns. This was a signal that brains react to an external stimulus by following their own internal dynamics, the stimulus acting solely as trigger.

G. Vitiello (✉)
Dipartimento di Fisica "E.R. Caianiello", Università di Salerno, and Istituto Nazionale di Fisica Nucleare, Fisciano, Salerno, Italy
e-mail: vitiello@sa.infn.it

J. Acacio de Barros, C. Montemayor (eds.), *Quanta and Mind*, Synthese Library 414, https://doi.org/10.1007/978-3-030-21908-6_9

In the activity of neocortex, the AM domains of synchronized oscillations form in few milliseconds (ms), have duration of 80–120 ms and carrier frequencies in the range of 12–80 Hz (beta-gamma range). They re-synchronize in frames at frame rates in the theta-alpha range (3–12 Hz) through a sequence of phase transitions. These patterns of oscillations extend over domains of linear size of 19 cm in humans, cover much of the hemisphere in rabbits and cats and are present in subjects engaged in cognitive interaction with environment, or else in resting, awake subjects; they are described as properties of the background activity of brains modulated in the engagement with the environment. Neither the electric field of the extracellular dendritic current nor the extracellular magnetic field from the high-density electric current inside the dendritic shafts, which are much too weak, nor the chemical diffusion, which is much too slow, appear to be able to fully account for the observed cortical collective activity (Capolupo et al. 2013; Freeman 2005; Freeman and Vitiello 2006).

Spontaneous breakdown of Symmetry in QFT. In the same years, in the decades of '60s–'80s, physicists were working at the foundation of the Standard Model of elementary particles and at the problem of the formation of ordered patterns in condensed matter physics, in systems such as superconductors, ferromagnets, crystals, etc. (Blasone et al. 2011; Bogoliubov et al. 1975; Umezawa 1993). The available theoretical tools were the quantum field theory (QFT) formalism, S-matrix theory, group theory. Soon after the formulation of the Dirac equation, in 1928, aimed to extend quantum mechanics (QM) to the domain of relativistic physics, it was realized that Dirac equation is actually an equation for systems with infinitely many degrees of freedom, thus not the equation for, e.g., the electron, but the equation for the electron *field*. Fields carry, by their same definition, infinitely many degrees of freedom. This is in contrast with Schrödinger equation, which describes one particle (or more than one particle, but *not* infinitely many particles). QM gives way to QFT and new horizons open to the theoretical analysis allowing the possibility to explore territories previously not accessible to QM. One crucial aspect of such a revolutionary change of paradigm was, for example, related with the Stone-von Neumann theorem in QM (von Neumann 1955), stating the *unitary equivalence* of all the representations of the canonical commutation relations (CCR) (or anti-commutation relations (CACR)). The theorem basic hypothesis of finite number of degrees of freedom fails to be true for fields and therefore the Stone-von Neumann theorem does not hold in QFT, where infinitely many *unitarily, i.e. physically, inequivalent* representations of the CCR (or CACR) therefore exist. The richness and the novelty of QFT, and its *dramatic* difference with respect to QM, originates from such exquisite mathematical feature (Blasone et al. 2011; Bogoliubov et al. 1975; Umezawa 1993).

In QM the *dogma of unitarity* keeps us confined *within* one single representation (or a class of unitarily equivalent representations). In QFT such a dogma may be released, which allows to study the whole manifold of the possible *physically inequivalent* dynamical regimes, or *phases,* in which the system can live and the *critical* processes of going through these representations, *phase transition* processes, can be also explored.

A collection of atoms, for example, may leave in the phase of atomic gas or in the crystal phase; in both the dynamical regimes, they are the "same atoms"; however, their dynamics manifests itself to us in *different physical behaviors*. A collection of elementary spin components (e.g. electrons) may be in the phase of a spin gas, or of a magnet, and so on with other examples.

The different phases present different *ordered patterns*, from the fully symmetric one (e.g. the rotational symmetry, without any preferred direction in the case of gas of spins) to the ordered one (the preferred direction of the magnetization signaling the preferred direction along which the majority of spins points in the average). These different phases are described by unitarily inequivalent representations in QFT, a possibility that is not allowed in QM. The study of a whole world of phenomena becomes now possible, from condensed matter physics, elementary particle physics to cosmology. For example, we live today in the *color confinement phase* of the Universe, where quarks, carrying the color quantum number, are indeed confined within the protons and other hadrons, and not in the primordial phase where quarks were unconfined.

In QFT the mechanism of *spontaneous breakdown of symmetry* (SBS) becomes possible, namely the possibility that a stimulus external to the system may trigger the transition from, say, the symmetric phase to an ordered one, e.g. from the spin gas to the magnetic phase. A weak external magnetic field may trigger indeed such a phase transition in the spin system. Once the trigger has operated, it may be switched off. The system follows its inner dynamics in reaching its new phase.

According to the Goldstone theorem (Blasone et al. 2011; Bogoliubov et al. 1975; Umezawa 1993), SBS generates in the system elementary components long range correlation waves, which are responsible of the system ordering. In quantum theories, to a wave it is always possible to associate a quantum particle or mode. The boson particles associated to our long range correlations are called the Nambu-Goldstone (NG) particles. They are massless bosons and therefore are able to span the whole system. The NG bosons are coherently condensed (Bose-Einstein condensation) in the system lowest energy state (the ground state or the vacuum). The vacuum condensate density provides a measure of the degree of ordering or coherence, described by the order parameter, a field specifying (labeling) the observed ordered pattern. Examples of NG modes are phonons in crystals, magnons (spin-wave quanta) in magnets, Cooper pairs in superconductors, etc.[1] (Alfinito et al. 2001; Blasone et al. 2011; Umezawa 1993).

Notice that ordering originates from the symmetry breakdown. *Order is lack of symmetry*. If symmetry is restored, order is lost (Blasone et al. 2011; Umezawa 1993). Also notice that in the formation of ordered patterns there is *transition from microscopic* (e.g. fully symmetric) behavior *to the macroscopic* behavior where the

[1]De-coherence in QM does not affect QFT coherence, which appears in a wide range of temperature, e.g. the diamond crystal loses its coherence (it melts) at a temperature of about +3545 °C; the common kitchen salt NaCl melts at +804 °C; the iron coherence of the elementary magnets is lost at 770 °C, in cobalt at +1075 °C. In superconductors, the critical temperatures in some niobium compound is about −252 °C; for some high-T_c superconductors it is a little above −153 °C.

elementary components, bound together by the ordering long range correlations, behave as *a whole*. In magnets, for example, the macroscopic behavior is described by the magnetization, called the order parameter, which is a *classical* field, namely not affected by quantum fluctuations at the elementary component level due to the system coherent state structure. *Coherence* reveals to be the central feature of the ordered pattern formation.

In SBS *macroscopic stable functions are thus dynamically generated out of the quantum dynamics of fluctuating elementary components. Macroscopic quantum systems* are obtained, *not* as classical limit of the quantum dynamics, but as truly dynamical result out of the quantum behavior of the elementary components.

The many-body brain model. I suppose that there is no need of much thinking to understand why Umezawa writes that (Umezawa 1995): "... In any material in condensed matter physics any particular information is carried by certain ordered pattern maintained by certain long range correlation mediated by massless quanta. It looked to me that this is the only way to memorize some information; memory is a printed pattern of order supported by long range correlations..." The QFT formalism was applied to brain modeling by Ricciardi and Umezawa (RU) in their 1967 paper *Brain and physics of many-body problems* (Ricciardi and Umezawa 1967), where they were observing that "... in the case of natural brains, it might be pure optimism to hope to determine the numerical values for the coupling coefficients and the thresholds of all neurons by means of anatomical or physiological methods.... First of all, at which level should the brain be studied and described? In other words, is it essential to know the behavior in time of any single neuron in order to understand the behavior of natural brains? Probably the answer is negative. The behavior of any single neuron should not be significant for the functioning of the whole brain, otherwise higher and higher degree of malfunctioning should be observed".

The RU effort was motivated by the still unanswered Lashely dilemma (Lashley 1948): "... Here is the dilemma. Nerve impulses are transmitted...form cell to cell through definite intercellular connections. Yet, all behavior seems to be determined by masses of excitation... within general fields of activity, without regard to particular nerve cells... What sort of nervous organization might be capable of responding to a pattern of excitation without limited specialized path of conduction? The problem is almost universal in the activity of the nervous system."

In the RU model of brain (Ricciardi and Umezawa 1967) (see also Atmanspacher 2015; Stuart et al. 1978, 1979) long range correlations are dynamically generated through the mechanism of SBS triggered by an external stimuli. In brains the water matrix is more than 80% of brain mass and it is therefore expected to be a major facilitator, mostly by ephapsis (Anastassiou et al. 2011; Freeman 2005), of neural interaction. The quantum variables are the electrical dipole vibrational field of biomolecules and water molecules. Notice that the neuron and the glia cells and other physiological units are classical, not quantum objects in the many-body model of brain.

The spontaneous breakdown of the rotational symmetry of the electrical dipole vibrational field dynamically generates the NG quanta, named the dipole wave quanta (DWQ) (Del Giudice et al. 1983, 1985, 1986, 1988; Jibu and Yasue 1995; Vitiello 1995, 2001). Their coherent condensation represents the memory, which is thus a "printed pattern of order supported by long range correlations", in Umezawa's words (see above and Umezawa 1995).

In the model, the recall of recorded memory occurs under a stimulus, "similar" to the one responsible for the memory recording, able to excite DWQ out of the ground state. Similarity between stimuli therefore refers not to their intrinsic features, but to the reaction of the brain to them; namely, to the possibility that under the stimuli action DWQ are condensed into, or excited from the ground state carrying the same ordering label.

There are, however, few shortcomings of the RU model. One is related with the memory capacity problem: any subsequent stimulus triggering the "new" SBS process cancels the previously recorded memory, thus overprinting the new memory over the previous one. Moreover, the model fails in explaining the observed coexistence of AM patterns and their irreversible time evolution.

The dissipative quantum model. It has to be stressed that brains are open, dissipative systems permanently exchanging energy, in various forms and modalities, with their environment. The RU model does not fully consider this crucial point, and it has been therefore extended to the dissipative dynamics, thus leading to "the dissipative quantum model of brain" (Pessa and Vitiello 2003, 2004; Vitiello 1995, 2001, 2009, 2014).

In dissipative QFT (Blasone et al. 2011; Celeghini et al. 1992) mathematics requires the doubling of the system degrees of freedom in order to represent the environment. This is described as the time-reversed copy of the system, its *Double*, since the energy flux outgoing from the system is incoming into the environment, and vice-versa. In the dissipative model, brain is thus in a continual dynamic interaction with its *Double* (Pessa and Vitiello 2003, 2004; Vitiello 1995, 2001, 2009, 2014).

The following chain of steps/processes schematically describes the dissipative many-body model: dissipation $\Rightarrow$ doubling of the system degrees of freedom, $A \rightarrow (A, \tilde{A})$, with $\tilde{A}$ denoting the "time-reversed mirror image" or "doubled modes" (the Double); external stimulus $\Rightarrow$ SBS $\Rightarrow$ dynamical generation of DWQ $\Rightarrow$ their condensation in a state labelled by the condensation density N (the order parameter identifying the memory code) $\Rightarrow$ time-evolution of the brain state $|0(t)>_N$ controlled by the energy flux balance, $E_0 = E_{Syst} - E_{Env} = 0$, and entropy variations $\Rightarrow$ free energy F minimization, $dF = 0$, $\Rightarrow$ irreversibility of time-evolution (breakdown of time-reversal symmetry), arrow of time (privileged direction in time-evolution).

The states $|0(t)>_N$, for any recorded memory code N, form the brain "memory space", or the brain state space (the space of unitarily inequivalent representations $\{|0(t)>_N\}$, each one associated to a memory code N). $|0(t)>_N$, for given N, describes a physical phase of the system identified by that specific N. The system may shift, under the influence of one or more stimuli acting as a control parameter, from phase to phase in the collection of brain-environment equilibrium vacua ($E_0 = 0$, $dF = 0$): the system undergoes an extremely rich sequence of phase transitions, i.e. a sequence of structures formed by AM patterns, as experimentally observed. The model indeed predicts (quasi-)non-interfering vacua (AM pattern textures), (phase) transitions among them (AM patterns sequencing), huge memory capacity, memories have a life-time (they can be forgotten). The "protection", or "not confusion" of memories is guaranteed by the unitarily inequivalence (orthogonality) of the representations $\{|0(t)>_N\}$, for each memory N (smoothing of the inequivalence, e.g. due to volume or impurity effects, may lead to transitions "from memory to memory").

The original many-body model could not describe these features. Predictions of the dissipative quantum model in agreement with experiments are here listed (Freeman and Vitiello 2010; Freeman et al. 2012):

- very low energy required to excite correlated neuronal patterns,
- AM patterns have large diameters, with respect to the small sizes of the component neurons,
- duration, size and power of AM patterns are decreasing functions of their carrier wave number k,
- there is lack of invariance of AM patterns with invariant stimuli,
- heat dissipation at (almost) constant in time temperature,
- the occurrence of spikes (vortices) in the process of phase transitions,
- the whole phenomenology of phase gradients and phase singularities in the vortices formation,
- the constancy of the phase field within the frames,
- the insurgence of a phase singularity associated with the abrupt decrease of the order parameter and the concomitant increase of spatial variance of the phase field (null spike),
- the onsets of vortices between frames, not within them,
- the occurrence of phase cones (spatial phase gradients) and random variation of sign (implosive and explosive) at the apex,
- the phase cone apices occur at random spatial locations,
- the apex is never initiated within frames, but between frames (during phase transitions).
- the fractal self-similarity and Bessel-like functional distribution of evoked potentials.
- The model leads to the classicality (not derived as the classical limit, but as a dynamical output) of functionally self-regulated and self-organized background activity of the brain.

In SBS theories, an important role is played by the nonlinearity of the dynamics, which also happens in the dissipative model. Let λ be the coupling constant measuring the intensity of the interaction among elementary components. One can show that in nonlinear dynamical systems, λ, although small, i.e. less than 1, enters in the solutions as its inverse $1/\lambda$. This means that the interaction cannot be switched off, i.e. never it can be $\lambda \to 0$, since never the denominator of a fraction can be set to zero. Weak intensity forces (small λ) may have then catastrophic consequences (proportional to $1/\lambda$), thus subverting the intuitive view that weak forces can be safely neglected. We have then non-perturbative physics where free fields or free systems cannot be defined. They are always interacting with other systems, e.g. with environment in which they are embedded. They are *open* or *dissipative* systems.

Meanings and mind. The emerging picture is that a stimulus selects a basin of attraction in the primary sensory cortex to which it converges (abstraction process), often with very little information as in weak scents, faint clicks, and weak flashes (trigger) (Freeman and Vitiello 2006; Vitiello 2004). The attractor selected by a stimulus is an instance of the category (generalization) that the attractor implements by its AM patterns: the waking state consists of a collection of potential states, any one of which but only one at a time can be realized through SBS (phase transition). The specific ordered pattern generated through SBS by an external input does not depend on the stimulus features, but on the system internal dynamics $\Rightarrow$ the stored memory is not a representation of the stimulus but of the *meaning* of the brain-environment interaction. *Memory is memory of meanings, not memory of information.*

The engagement of the subject with the environment in the *action-perception cycle* is the essential basis for the emergence and maintenance of *meaning* through successful interaction and its knowledge. The environment acts on the self independently as well as reactively. The "inter-action" (active mirror) is ruled by the free energy minimization processes.

The continual balancing of the energy fluxes at the brain–environment interface amounts to the continual updating of the meanings of the flows of exchanged information in the brain behavioral relation with the environment. By repeated trial-and-error, the brain constructs within itself an understanding of its surround, which constitutes its knowledge of its own world, that is its Double.

Brain activity is a process of transformation of information collected through perceptual inputs into meanings, analogous to the transition from the syntax to the semantics in linguistics (Piattelli-Palmarini and Vitiello 2015).

One of the merits of the dissipative model is the possibility of deriving from the coherent quantum dynamics the classicality of trajectories in the memory space. These trajectories are found to be classical deterministic chaotic trajectories (Pessa and Vitiello 2003, 2004; Vitiello 2004). The dissipative model also accounts for the fractal self-similarity in brain background activity (Vitiello 2009, 2012).

A crucial neural mechanism observed by Freeman in his laboratory (included in the above list of results) is that the event that initiates the transition to a perceptual state is an abrupt decrease to near zero in the analytic power of the background activity, called a null spike. It is associated with the increase of spatial variance of the analytic phase, occurring aperiodically at rates in the theta (3–7 Hz) and alpha (8–12 Hz) ranges. It is called a vortex since it has rotational energy at the geometric mean frequency of the pass band. The vortex occupies the whole area of the phase-locked neural activity of the cortex for a point in time.

Between null spikes the cortical dynamics is (nearly) stationary for $\sim$60–160 ms. This is called a frame. During periods of high amplitude the spatial deviation of phase is low, the phase spatial mean tends to be constant within frames and changes suddenly between frames. The reduction in the amplitude of the spontaneous background activity induces a brief state of indeterminacy in which the significant pass band of the electrocorticogram (ECoG) is near to zero and the phase of ECoG is undefined.

Each null spike initiates a spatial phase cone, namely a spatial phase gradient that is imposed on the carrier wave of the AM pattern in a frame. The arriving stimulus can drive the cortex across a phase transition process to a new AM pattern. The observed velocity of spread of phase transition is finite (there is no "instantaneous" phase transition). These features have been documented as markers of the interface between microscopic and mesoscopic phenomena. Observation shows that the location of the phase cone apex is a random variable across frames, determined by the accidental site where the null spike is lowest and the background input is highest.

The model prediction is in agreement with these observations: the initial site where non-homogeneous condensation starts (the phase cone apex) is not conditioned by the incoming stimulus, but is randomly determined by a number of concurrent local conditions. The apex is never initiated within frames (in the broken symmetry phase or ordered region), but between frames (during phase transitions). From each frame to the next, it is observed the random variation of the slope of the conic phase gradient, negative with explosion, positive with implosion (Freeman and Vitiello 2010; Freeman et al. 2012). These two regimes are actually described by retarded and advanced Green's functions (forward and backward in time) and give neuronal consistency to the Double.

Laboratory observations thus show that time-reversed copies of the AM patterns are formed in the brain (Freeman and Quian Quiroga 2013; Freeman 1975/2004, 2014; Freeman and Vitiello 2016; Kay and Freeman 1998; Vitiello 2018). The Double forms 'reversed in time' copies of the AM patterns and it appears to be the necessary reference for the formation of meanings generated by the perceptual experience. The time-reversed copies are "built in" in the brain dynamics. The observed Bessel-like functional distributions for average evoked potential data can be indeed analyzed in the dissipative model in terms of a couple of damped and amplified oscillators (forward and backward in time) (Capolupo et al. 2017; Freeman et al. 2015). In this sense, the brain dissipative activity is dynamically "self-generative", the action-perception cycle contains in itself a dynamic "truth-evaluation function", a control mechanism discriminating between two similar behaviors or perceptions, continuously tested and dynamically re-modeled in the optimization of "to-be-in-the-world" of the subject (Capolupo et al. 2017; Freeman et al. 2015). Quoting from (Freeman and Vitiello 2016), "the Double is mind, not matter, yet it is completely entangled with matter, the AM pattern. The crucial point is that activity *cannot* be separated as mental activity versus brain activity. They are intrinsically related, *dynamically entangled* in a precise formal way shown by the many-body model. In such an entanglement might reside the act of consciousness.... The relation between matter and mind is thus expressed in terms of dynamic trajectories along parallel time lines, one corresponding to reconstructing the past in remembering, the other to forecasting environmental trends by extrapolation into the future in predicting."

In conclusion, the scenario offered by QFT and the dissipative quantum model point to a vision in agreement with some of the Schrödinger's observations in his *What is life?* (Schrödinger 1944), where talking of biological systems in general, but his remarks well apply also to brains, he says that "regularities only in the average" (p. 78), as the ones derivable by use of the "statistical mechanisms" cannot explain the "enigmatic biological stability" (p. 47). In fact, "... it needs no poetical imagination but only clear and sober scientific reflection to recognize that we are here obviously faced with events whose regular and lawful unfolding is guided by a "mechanism" entirely different from the "probability mechanism" of physics" (p. 79) and the attempt to explain the biological functional stability in terms of the regularities of statistical origin would be the "classical physicist's expectation" that "far from being trivial, is wrong" (p. 19). On the other hand, in his book *The Computer and the Brain* (von Neumann 1958), von Neumann observes that "... the mathematical or logical language truly used by the central nervous system is characterized by less logical and arithmetical depth than what we are normally used to. ... We require exquisite numerical precision over many logical steps to achieve what brains accomplish in very few short steps".

Let me close with the following passage by Jorge Louis Borges on the self and its Double (Borges 1960):

> The other one, the one called Borges, is the one things happen to.... It would be an exaggeration to say that ours is a hostile relationship; I live, let myself go on living, so that Borges may contrive his literature, and this literature justifies me.... Besides, I am destined to perish, definitively, and only some instant of myself can survive him.... Spinoza knew that all things long to persist in their being; the stone eternally wants to be a stone and a tiger a tiger. I shall remain in Borges, not in myself (if it is true that I am someone).... Years ago I tried to free myself from him and went from the mythologies of the suburbs to the games with time and infinity, but those games belong to Borges now and I shall have to imagine other things. Thus my life is a flight and I lose everything and everything belongs to oblivion, or to him.
>
> I do not know which of us has written this page.

References

Alfinito, E., Viglione, E. G., & Vitiello, G. (2001). The decoherence criterion. *Modern Physics Letters, B15*, 127–135.

Anastassiou, C. A., Perin, R., Markram, H., & Koch, C. (2011). Ephaptic coupling of cortical neurons. *Nature Neuroscience, 14*, 217–223. https://doi.org/10.1038/nn.2727/.

Atmanspacher, H. (2015). Quantum approaches to consciousness. *Stanford Encyclopedia of Philosophy*. http://plato stanford edu/entries/qt-consciousness/.

Blasone, M., Jizba, P., & Vitiello, G. (2011). *Quantum field theory and its macroscopic manifestations*. London: Imperial College Press.

Bogoliubov, N. N., Logunov, A., & Todorov, I. (1975). *Introduction to axiomatic quantum field theory*. Reading: Benjamin.

Borges, J. L. (1960). Borges and I. In *El hacedor*. Biblioteca Borges Alianza Editorial.

Capolupo, A., Freeman, W. J., & Vitiello, G. (2013). Dissipation of 'dark energy' by cortex in knowledge retrieval. *Physics of Life Reviews, 10*(1), 85–94.

Capolupo, A., Kozma, R., Olivares del Campo, A., & Vitiello, G. (2017). Bessel-like functional distributions in brain average evoked potentials. *Journal of Integrative Neuroscience, 16*, S85–S98.

Celeghini, E., Rasetti, M., & Vitiello, G. (1992). Quantum dissipation. *Annals of Physics, 215*, 156–170.

Del Giudice, E., Doglia, D., Milani, M., & Vitiello, G. (1983). Spontaneous breakdown of symmetry and boson condensation in biology. *Physics Letters, 95A*, 508–510.

Del Giudice, E., Doglia, D., Milani, M., & Vitiello, G. (1985). A quantum field theoretical approach to the collective behavior of biological systems. *Nuclear Physics B, 251*(FS13), 375–400.

Del Giudice, E., Doglia, D., Milani, M., & Vitiello, G. (1986). Electromagnetic field and spontaneous symmetry breaking in biological matter. *Nuclear Physics B, 275*(FS 17), 185–199.

Del Giudice, E., Preparata, G., & Vitiello, G. (1988). Water as a free electric dipole laser. *Physical Review Letters, 61*(9), 1085–1088.

Freeman, W. J. (1975/2004). *Mass action in the nervous system*. New York: Academic.

Freeman, W. J. (1991). The physiology of perception. *Scientific American, 264*, 78–85.

Freeman, W. J (2005). NDN, volume transmission, and self-organization in brain dynamics. Journal *of Integrative Neuroscience, 4*(4), 407–421.

Freeman, W. J. (2014). Mechanism and significance of global coherence in scalp EEG. *Current Opinion in Neurobiology, 31*, 199–205.

Freeman, W. J., & Quian Quiroga, R. (2013). *Imaging brain function with EEG*. New York: Springer.

Freeman, W. J., & Vitiello, G. (2006). Nonlinear brain dynamics as macroscopic manifestation of underlying many-body dynamics. *Physics of Life Reviews, 3*, 93–117.

Freeman, W. J., & Vitiello, G. (2010). Vortices in brain waves. *International Journal of Modern Physics, B23*, 3269–3295.

Freeman, W. J., & Vitiello, G. (2016). Matter and mind are entangled in two streams of images guiding behavior and informing the subject through awareness. *Mind and Matter, 14*(1), 7–24.

Freeman, W. J., Livi, R., Obinata, M., & Vitiello, G. (2012). Cortical phase transitions, non-equilibrium thermodynamics and the time-dependent Ginzburg-Landau equation. *International Journal of Modern Physics, B26*(6), 1250035.

Freeman, W. J., Capolupo, A., Kozma, R., Olivares del Campo, A., & Vitiello, G. (2015). Bessel functions in mass action modeling of memories and remembrances. *Physics Letters, A379*, 2198–2208.

Jibu, M., & Yasue, K. (1995). *Quantum brain dynamics and consciousness*. Amsterdam: John Benjamins.

Kay, L. M., & Freeman, W. J. (1998). Bidirectional processing in the olfactory-limbic axis during olfactory behavior. *Behavioral Neuroscience, 112*(3), 541–553.

Lashley, K. (1948). *The mechanism of vision* (pp. 302–306). Provincetown: Journal Press.

Pessa, E., & Vitiello, G. (2003). Quantum noise, entanglement and chaos in the quantum field theory of mind/brain states. *Mind and Matter, 1*, 59–79.

Pessa, E., & Vitiello, G. (2004). Quantum noise induced entanglement and chaos in the dissipative quantum model of brain. *International Journal of Modern Physics, B18*, 841–858.

Piattelli-Palmarini, M., & Vitiello, G. (2015). Linguistics and some aspects of its underlying dynamics. *Biolinguistics, 9*, 96–115.

Ricciardi, L. M., & Umezawa, H. (1967). Brain and physics of many-body problems. *Kybernetik, 4*, 44–48; Reprinted in G. Globus, K. H. Pribram, G., & Vitiello (Eds.) (2004). *Brain and being* (pp. 255–266). Amsterdam: John Benjamins Publ Co.

Schrödinger, E. (1944). *What is life?* Cambridge: Cambridge University Press.

Stuart, C. I. J., Takahashi, Y., & Umezawa, H. (1978). On the stability and non-local properties of memory. *Journal of Theoretical Biology, 71*, 605–618.

Stuart, C. I. J., Takahashi, Y., & Umezawa, H. (1979). Mixed system brain dynamics: Neural memory as a macroscopic ordered state. *Foundations of Physics, 9*, 301–327.

Umezawa, H. (1993). *Advanced field theory: Micro macro and thermal concepts*. New York: American Institute of Physics.

Umezawa, H. (1995). Development in concepts in quantum field theory in half century. *Mathematical Japonica, 41*, 109–124.
Vitiello, G. (1995). Dissipation and memory capacity in the quantum brain model. *International Journal of Modern Physics, B9*, 973–989.
Vitiello, G. (2001). *My double unveiled*. Amsterdam: John Benjamins Publ Co.
Vitiello, G. (2004). The dissipative brain. In G. Globus, K. H. Pribram, & G. Vitiello (Eds.), *Brain and being. At the boundary between science philosophy language and arts* (pp. 315–334). Amsterdam: John Benjamins Publ Co.
Vitiello, G. (2009). Coherent states fractals and brain waves. *New Mathematics and Natural Computation, 5*, 245–264.
Vitiello, G. (2012). Fractals coherent states and self-similarity induced noncommutative geometry. *Physics Letters A, 376*, 2527–2532.
Vitiello, G. (2014). The use of many-body physics and thermodynamics to describe the dynamics of rhythmic generators in sensory cortices engaged in memory and learning. *Current Opinion in Neurobiology, 31*, 7–12.
Vitiello, G. (2018). Brain and its mindful double. *Journal of Consciousness Studies, 25*(1–2), 151–176.
von Neumann, J. (1955). *Mathematical foundation of quantum mechanics*. Princeton: Princeton University Press.
von Neumann, J. (1958). *The computer and the brain* (pp. 80–81). New Haven: Yale University Press.

Part II
Mind Informs Quanta

Chapter 10
Contextuality Revisited: Signaling May Differ From Communicating

Harald Atmanspacher and Thomas Filk

Abstract The notion of contextuality in quantum theory expresses that the result of a measurement (e.g. performed by Alice) depends on the experimental context or, more precisely, on other measurements (e.g. performed by Bob). This kind of contextuality presupposes that signals transferring information about Bob's experiment to Alice (and vice versa) are excluded. In quantum physics this can be guaranteed if the two measurements are performed within the causal complements of their lightcones. In this case, signaling would violate special relativity. Some recent scenarios in cognitive science apply a similar non-signaling condition to test whether measurements on cognitive systems are contextual. For a refined discussion of contextuality in such scenarios, we argue that it is important to distinguish two types of signaling: (1) signaling that Alice and Bob can use to communicate, and (2) signaling that Alice and Bob cannot use to communicate. Lacking communication may be inconclusive for contextuality if signaling without communication is still present.

10.1 Introduction

A violation of Bell inequalities (Bell 1966) is generally considered to be the lithmus test for experimental evidence of quantum effects. In the realm of physics the violation of Bell-type inequalities (the so-called CHSH inequalities) has been proven experimentally without any reasonable doubt by the seminal experiments of Aspect et al. (1982).

H. Atmanspacher (✉)
Collegium Helveticum, University of Zurich and ETH Zurich, Zürich, Switzerland
e-mail: atmanspacher@collegium.ethz.ch

T. Filk
Institute for Physics, University of Freiburg, Freiburg im Breisgau, Germany

Parmenides Foundation for the Study of Thinking, Munich, Germany

J. Acacio de Barros, C. Montemayor (eds.), *Quanta and Mind*, Synthese Library 414, https://doi.org/10.1007/978-3-030-21908-6_10

In recent years, the formalism of quantum theory has also been successfully applied to various scenarios of cognitive science, such as decision processes (Pothos and Busemeyer 2013), similarity judgements (Aerts et al. 2011; Pothos et al. 2013), order effects (Atmanspacher and Römer 2012; Wang et al. 2014), bistable perception (Atmanspacher and Filk 2013a), learning on networks (Atmanspacher and Filk 2006), meaning in natural language (Bruza et al. 2015), and behavioral economy (Haven and Khrennikov 2013). Therefore, it is a natural question to look for violations of Bell-type inequalities in these scenarios as well.

However, there are many possible loopholes that may undercut the consequence that violating Bell-type inequalities implies contextuality. One of these loopholes is that a measurement performed by Alice may impose an influence – for instance in the form of a signal – onto the measurement performed by Bob. In the experiments of Aspect this loophole was closed by performing the measurements within the light-cone complements of each other. If we assume that signals cannot propagate faster than with the speed of light, then signaling becomes impossible in such a setup.

In cognitive scenarios it is much harder to exlude this loophole. In fact, Dzhafarov and collaborators (see, e.g., Kujala and Dzhafarov 2016; Dzhafarov et al. 2016) have pointed out that in most cognitive experiments the violation of Bell-type inequalities vanishes if signaling between two stimuli is properly taken into account. As a consequence, contextuality can be explained by a direct influence of the first measurement onto the result of the second one.

In a recent paper, Cervantes and Dzhafarov (2018) reported results of a psychological experiment in which a violation of Bell-type inequalities can be shown to persist even after the influence of possible signaling processes has been taken into account and subtracted. To be precise, they applied a particular "non-signaling" criterion which constitutes a measure of the extent to which correlations can be used by Bob and Alice to exchange information (to communicate).

In the following we will argue that the non-signaling criterion applied by Cervantes and Dzhafarov (2018) is useful to establish evidence for an influence between two measurements that can be used to communicate between them. However, the criterion is not sufficient to exclude influences between them that cannot be used for communication. To make this more transparent, we propose to distinguish between (1) signals or influences that allow the (intentional) transfer of information and (2) signals or influences that do not allow such transfer of information (see, e.g., Filk 2015, 2016; Walleczek and Grössing 2016).

In the next section, we will briefly review (Bell-type) CHSH inequalities. In Sect. 10.3 we will introduce the non-signaling condition used by Cervantes and Dzhafarov (2018). Section 10.4 elaborates why this criterion may be necessary but not sufficient in order to exclude direct influences from one experiment to another. Section 10.5 will give some explicit examples of systems which violate the CHSH inequality maximally even when signaling is taken into account à la Dzhafarov and colleagues. The article ends with a brief summary.

10.2 Bell-Type Inequalities and Contextuality in Physics

We consider the following situation: Alice and Bob each have a measuring device for some binary property (i.e. the possible measurement results can only be $+1$ and -1) of a shared physical system. A typical shared physical system is a particle pair where Bob has access to one particle and Alice to the other, and both Alice and Bob can measure the properties of their particle. To be concrete, we might consider an entangled two-photon state and the measurements refer to the measurements of the polarization (for simplicity we only consider linear polarizations characterized by the angle of a polarization axis) of the photons. These polarization measurements can be performed with polarizing filters or polarizing beam-splitters followed by photon detectors.

Let us now suppose that Alice and Bob have two choices for the orientation of their beam-splitters. Alice can set her beam-splitter to an angle α_1 (which defines the two possible polarization axes of a transmitted photon) or an angle α_2. If the photon is detected in the transmitted directions, the result of the measurement is $+1$, if it is detected in the reflected direction, the result is -1. Similarly, Bob can choose for the axis of his beam-splitter the angles β_1 or β_2 with the same rules for the results of the measurements. Alice and Bob each can only perform one measurement on their photon, yielding one of two values which we call A_i for Alice's settings and B_i for Bob's. $\langle A_i B_j \rangle$ denotes the expectation value (averaged over many repeated measurements on many identically prepared photon pairs) of the product of the two quantities.

In the derivation of Bell-type inequalities – we will use the CHSH version (after Clauser et al. 1969) – one has to assume that the probability distributions for the results obtained by Alice are independent of Bob's settings and vice versa. With this assumption one can show that

$$\left| \left(\sum_{i,j} \langle A_i B_j \rangle \right) - 2 \max_{i,j} \langle A_i B_j \rangle \right| \leq 2\,. \tag{10.1}$$

In a simplified form this equation tells us that it is not possible to have $A_1 = B_1 = A_2 = B_2$ but $A_1 \neq B_2$ (or similar logical contradictions), if all four quantities do objectively exist in the sense that they are "elements of reality". If this were the case the quantity on the left hand side of (10.1) would assume the value 4, the maximal value theoretically possible. A system which yields this value is sometimes called a PR-box, after Popescu and Rohrlich (2004). For quantum systems it can be shown that (10.1) can be violated for certain settings of the beam-splitters, i.e. for certain values of α_i and β_j. However, the maximal possible violation in quantum systems is given by Tsirelson's bound of $2\sqrt{2}$ (Tsirelson 1980).

Any violation of (10.1) can easily be explained if Alice and Bob can "communicate", i.e. influence each other directly after the first measurement has been performed. For instance, a simple way to violate (10.1) can be obtained if the following (hypothetical) rules are assumed:

1. Whenever Alice's setting is α_1, Alice and Bob will always get the same result (both $+1$ or both -1) independent of Bob's setting.
2. When Alice's setting is α_2, Bob will always get the same result as Alice if his setting is β_2, otherwise (if his setting is β_1) he will obtain the result opposite to Alice's.

Under these rules we obtain $\langle A_1B_1\rangle = \langle A_1B_2\rangle = \langle A_2B_2\rangle = -\langle A_2B_1\rangle$ and the left hand side of (10.1) is 4. However, the information about Alice's choice (α_1 or α_2) has to be transmitted to Bob to make this work. On the other hand, if Bob makes his measurement before Alice, the information about his setting has to be transmitted to Alice. In Sect. 10.4 we will discuss in detail the kinds of information (about the setting and/or about the results of the measurements) that can be transmitted.

Violations of Bell-type inequalities are related to the notion of contextuality. This notion historically derives from a theorem by Kochen and Specker (1967). It proves that in general (for Hilbert spaces with dimension ≥ 3) it is not possible so assign a definite value to some observable A independent of which other observable B_1 or B_2 is measured. While A and B_i may commute and thus can be measured with arbitrary precision simultaneously, B_1 and B_2 do not commute and define different contexts for measuring A.

In its original version this theorem makes no reference to different (perhaps far remote) sites of an entangled physical system, and signaling was not an issue. However, the remarkable consequences of quantum theory become most prominent if A refers to an observable measured on one particle, while B_i are observables which can be measured on a remote second particle. B_1 and B_2 cannot be arbitrarily precisely determined together because they do not commute, but each of them can be arbitrarily precisely determined together with A. In this way, the CHSH inequalities can be formulated in terms of contextuality, and in this case one can prove that the correlations between the measured values cannot be used for signaling.

10.3 The Non-signaling Criterion by Dzhafarov and Colleagues

Apart from entangled quantum systems in physics, Bell-type inequalities have been employed for experiments in the cognitive sciences as well. The general paradigm for such experiments can be roughly described as follows: A group of subjects is divided into four subgroups defining four contexts (this corresponds to the four possible settings (α_i, β_j), $i, j = 1, 2$) and each subject in its subgroup has to make two choices (corresponding to the answers A_i and B_j to be ± 1) for the two pairs of alternatives defining their context.

Several groups reported violations of Bell-type inequalities in such experiments (e.g., Aerts 2014; Aerts et al. 2013; Bruza et al. 2015), thus supporting the idea that certain cognitive experiments need a quantum formalism (in particular a Hilbert-space formalism) to explain these violations by entangled cognitive states. The idea is that violations of Bell-type inequalities even in cognitive systems cannot be described classically and are associated with the notion of quantum contextuality.

However, the reported correlations rarely seem as counterintuitive as they first appeared in genuine quantum systems. On the contrary, correlations between cognitive states typically look quite plausible, and standard reactions to quantum-type explanations were often accompanied by some uneasiness as to the invocation of quantum entanglement as the best or only explanation. Soon it became clear that the condition of non-invasiveness (Atmanspacher and Filk 2013b) or "non-signaling" of one measurement onto the results of the other is difficult to realize.

In the presence of signaling, it is easy to construct classical systems which violate Bell-type inequalities. In a series of papers, Dzhafarov and colleagues (see Kujala and Dzhafarov 2016; Dzhafarov et al. 2016) found a clever way to distinguish correlations due to signaling from non-signaling correlations in such situations. If the signaling component is taken into account and subtracted in Bell-type (e.g. CHSH) inequalities, it becomes possible to study whether the remaining correlations still violate the inequality and, thus, exhibit "genuine" contextuality.

Essentially their definition of signaling refers to the conditional expectation values of the measurement results of Alice and Bob, depending on their mutual settings. If the expectation values of Alice for A_i depend on the settings of Bob's measurements, i.e. whether he has measured (or is about to measure) β_1 or β_2, and/or vice versa, there is signaling in the sense that Alice and Bob are capable of exchanging information.

Such signaling is defined as the sum over the differences between the expectation values for different settings (Dzhafarov et al. 2016, Eq. 26):

$$I_S = \sum_{i=1,2} \left| \langle A_i \rangle_1 - \langle A_i \rangle_2 \right| + \sum_{j=1,2} \left| \langle B_j \rangle_1 - \langle B_j \rangle_2 \right| \qquad (10.2)$$

where $\langle \cdot \rangle_k$ $(k = 1, 2)$ in the case of the variables A_i corresponds to Bob's settings β_k and in the case of the variables B_j to Alice's settings α_k $(k = 1, 2)$.[1] Now Dzhafarov and colleagues replace the CHSH inequality (10.1) by (cf. Dzhafarov et al. 2016, theorem 7.2):

[1]We prefer to use a notion more familiar in physics where the results of measurements are not denoted as random variables (as by Dzhafarov and colleagues) but as observables for which the measured results can be subject to different distributions. Of course, the meaning is the same as, e.g., in the notation $\langle A_i \rangle_k \equiv E[A_i^k]$ used by Dzhafarov and colleagues.

Table 10.1 Three possible scenarios for settings and results with no room for randomness: the results are fixed once the settings are fixed. All scenarios yield a maximal violation of the CHSH inequality. Left: Alice and Bob can both exchange information. If Alice wants to send a message to Bob, Bob puts his setting to β_2 and Alice switches between α_1 and α_2 to send the bits $+1$ and -1, respectively. Vice versa, Bob can use his switch β_2 to send controlled signals to Alice if she puts her setting to α_2. Middle: Alice can send information to Bob but not vice versa, if Bob sets his setting to β_1. Right: Bob can send information to Alice but not vice versa, if she sets her setting to α_2

α	β	A	B
1	1	1	1
1	2	1	1
2	1	1	−1
2	2	−1	−1

α	β	A	B
1	1	1	1
1	2	1	1
2	1	1	−1
2	2	1	1

α	β	A	B
1	1	1	1
1	2	1	1
2	1	−1	1
2	2	1	1

$$\left| \left(\sum_{i,j=1,2} \langle A_i B_j \rangle \right) - 2\max_{i,j} \langle A_i B_j \rangle \right| - I_S \leq 2 \,. \tag{10.3}$$

The extent I_S to which information can be transferred between Alice and Bob is subtracted from the left hand term in (10.1), such that signaling contributions are corrected for. If this inequality is not violated, one must not speak of "genuine" contextuality.

Table 10.1 shows three examples with a maximal possibility for signaling under the assumption that all results are fixed once the settings are fixed. The examples also show that signaling is not a mutual or both-ways property. The contribution is $I_S = 6$ if both Alice and Bob can use the device for signaling (left scenario in Table 10.1), and it is $I_S = 2$ when either Bob or Alice can send messages (middle and right scenarios in Table 10.1). In all cases the inequality (10.3) is satisfied and there is no contextuality. Compare a similar example by Dzhafarov et al. (2016, Tab. 6).

10.4 Signaling Versus Communicating

Given the discussion of fixed relations between settings and results according to Table 10.1, the question arises whether by introducing correlated randomness one can decrease the possibility for signaling while keeping the violation of (10.1) maximal. This is possible indeed! The example mentioned in Sect. 10.2 is of this kind: The distributions for the random variables remain the same: they are equally distributed, i.e., both Alice and Bob see a distribution of all their results with a probability of 1/2, independent of the settings.

In order to elaborate on this point, Table 10.2 shows sets of (fictitious) probabilities p_{ij} for the occurrence of correlated pairs. Note that the CHSH-part, i.e. the contributions from non-signaling correlations to (10.3), does not change because the

Table 10.2 A scenario with probabilistic results (A_i, B_j). The left part of Table 10.1 is obtained for $p_{11} = p_{12} = p_{21} = p_{22} = 1$. The CHSH-part of inequality (10.3) remains unchanged and leads to a maximal violation of 4. However, the degree of signaling is changed

α	β	(1, 1)	(1, −1)	(−1, 1)	(−1, −1)
1	1	p_{11}	0	0	$(1 - p_{11})$
1	2	p_{12}	0	0	$(1 - p_{12})$
2	1	0	p_{21}	$(1 - p_{21})$	0
2	2	$(1 - p_{22})$	0	0	p_{22}

expectation values for the correlations do not change. However, the signaling part according to Eq. (10.2) and Table 10.2 now becomes:

$$I_S = |2p_{11} - 2p_{12}| + |2p_{21} + 2p_{22} - 2| + |2p_{11} + 2p_{21} - 2| + |2p_{12} + 2p_{22} - 2| \quad (10.4)$$

We immediately see that this signaling part vanishes, $I_S = 0$, if and only if all probabilities are $p_{ij} = 1/2$. Of course, one can introduce additional probabilities relaxing the strict correlations we have required so far. In this case also the CHSH-part will change and one obtains values closer to Tsirelson's bound of $2\sqrt{2}$.

Let us emphasize an important difference between the notions of "no direct influence" or "non-invasiveness" (Atmanspacher and Filk 2013b) and "marginal selectivity", as used by Dzhafarov and coworkers in previous publications (e.g., Dzhafarov 2003). In quantum information theory "marginal selectivity" is known from the "non-communication theorem" (Florig and Summers 1997).

Unfortunately, the term "non-signaling" has been used in both contexts. Walleczek and Grössing (2016) distinguish "non-signaling" (in the sense of no direct influence whatsoever) from "non-communicating" (in the sense that signaling may be present but cannot be used for communication). The examples in the next section are supposed to show that classical physical systems may violate the CHSH inquality maximally (like a PR-box) but the correlations cannot be used for communication. In our terminology there is "direct influence" or "invasiveness", but there is also "marginal selectivity" (or, in other words, $I_S = 0$).

10.5 Classical Systems Violating the CHSH Inequality Maximally

Is has been noted by Cervantes and Dzhafarov (2018) that it is easy to formulate questions which allow for "correct answers" so that the CHSH inequality is violated maximally. It is much harder to find questions for which the probability among a population of subjects to choose between the various possibilities is roughly 1/2. Such systems would be marginally selective and contextual in a probabilistic sense.

Filk (2015) constructed an example of a PR-box which cannot be used for communication as expressed by the criterion of Dzhafarov and colleagues. This example is a simple physical system that can be implemented as an electronic or even mechanical device and realizes Table 10.2 with all probabilities equal to 1/2. It can even be modified to the effect that the order in which Alice and Bob make their measurements does not matter. In this case the output of the later measurement will be retarded with respect to the settings and the output of the first measurement such that it will be located within the lightcone of that first measurement and does not violate special relativity. The overall conclusion is that a non-communicating (in the sense of Cervantes and Dzhafarov 2018) PR-box is physically possible, although there clearly is a direct influence between measurements.

Let us now consider the following scenario: The two experimental choices of Alice are realized by pairs of numbers $\alpha_1 = (2, 9)$ and $\alpha_2 = (3, 4)$ with the possible outcomes $A_1 = 2$ or 9 and $A_2 = 3$ or 4. The two experimental choices of Bob are number characteristics: $\beta_1 =$ (prime, non-prime) and $\beta_2 =$ (even, odd). The possible results are $B_1 =$ "prime" or "non-prime" and $B_2 =$ "even" or "odd". Table 10.3 summarizes these possibilities.[2]

This scenario is mathematically straightforward and does (in principle) not leave room for inappropriate interpretation. Moreover, the mathematical concepts are sufficiently neutral in order to avoid subjective preferences, thus raising the violation of the CHSH inequality to close to its maximal value of 4. If I_S turned out to be small enough, the remaining correlations from a corresponding experiment may not only violate the classical bound (as in Cervantes and Dzhafarov 2018) but also the Tsirelson bound of $2\sqrt{2}$. Such a case would render not only classical models but also conventional Hilbert-space models inappropriate. See also Carmi and Cohen (2018) for related discussion.

If correct assignments are given in all cases, the CHSH inequality is violated maximally (note that 2 is a prime number). The question is to which extent subjects have a preference for some of the numbers within the pairings. Such a preference would define the probabilities p_{ij} and, thereby, the extent to which the non-communicating condition is violated, leading to the correction term I_S of (10.3). The "signaling" between the first question (choosing a number) and the second (giving

Table 10.3 Each context consists of two choices, between two numbers and two attributes. If correct answers are given, the CHSH inequality is violated maximally. The assignment is such that the first line for each context is attributed a $+1$ and the second line a -1

	Number	Characteristic
Context 1	2	Prime
	9	Non-prime
Context 2	2	Even
	9	Odd
Context 3	3	Prime
	4	Non-prime
Context 4	3	Even
	4	Odd

[2]Note that Table 10.3 closely resembles the context table studied by Cervantes and Dzhafarov (2018) based on the characters from the fairy-tale "The Snow Queen" by Andersen.

a characteristic) is obvious: subjects know which pair of numbers defined the first part of the context and they know which choice was made before the second part of the context is given. Otherwise, a correct answer would not be possible.

Cervantes and Dzhafarov (2018) argue that their framework refers to mathematical properties of a system described by random variables and concede that it may disregard "hidden" signaling influences of physical or psychological origin. In this sense they conclude that their correction term I_S, which does not account for "hidden" signaling, is sufficient to decide about contextuality. However, physical intuition would entail that signaling whatsoever, "hidden" or not, cannot create true quantum contextuality. Otherwise, purely classical systems as the ones discussed here would be contextual although their behavior can plausibly be understood once the hidden signaling is recognized.

By contrast, a highly counterintuitive situation (in the face of special relativity) would result from the following scenario: Alice and Bob are far apart from each other in the company of scientists (it should be guaranteed that neither Alice nor Bob nor the scientists can communicate with one another). Alice's scientist chooses for her the left part of a context (one of the pairs $(2, 9)$ or $(3, 4)$), while Bob's scientist chooses for him the right part of a context (in his case the result will be either the pair (prime, non-prime) or the pair (even, odd)). Now Alice chooses between the two numbers assigned to her and Bob chooses between the two attributes assigned to him. If the CHSH inequality were violated in this scenario, this would be truly surprising – if not irritating. This were the kind of contextuality that would be required in order to legitimately conclude that Alice and Bob are entangled in a true quantum sense.

10.6 Summary

The feature of contextuality in quantum systems means that the result of a measurement depends on the details of the experiment, in particular on what one chooses to measure alongside with that measurement. There is a close relationship of contextuality with the violation of Bell-type inequalities, to the effect that entangled quantum systems exhibit both features. If the correlations violating Bell-type inequalities are due to signaling, the violation and its associated contextuality can be trivially explained in classical terms. Such signaling can be elegantly avoided in experiments whose results are located in the causal complements of each other's lightcone.

Contextuality and violations of Bell-type inequalities have recently been studied in cognitive systems as well. In such systems, signaling is much harder to exclude experimentally. In view of this difficulty, Dzhafarov and colleagues designed a measure I_S that serves the post hoc distinction of signaling versus non-signaling correlations in Bell-type inequalities. However, this measure does not explicitly distinguish between kinds of signaling that can or cannot be used to transfer information. In other words, I_S does not distinguish between "communicating" and "non-communicating" signals.

We constructed examples in which signals are exchanged between two measurements, but these signals cannot be used for a transfer of information from one to the other. The examples show that the distinction between communicating signals and non-communicating signals is significant. In cognitive experiments one may be able to exclude or correct for actual communication but it seems very difficult to operationally exclude signals that cannot be used for communication. In this case, the approach by Dzhafarov and colleagues would indicate "genuine" quantum-type contextuality based on the violation of their criterion for non-communication. From a quantum point of view, this violation could be considered inconclusive for contextuality if signaling without communication is present.

Note Added in Proofs

After the present article was submitted, Eric Cavalcanti (2018) published a paper suggesting a distinction similar to the one we draw between signaling that can be used for communication and "hidden" signaling that cannot. Cavalcanti discusses contextuality within the approach of classical causal models, where he contrasts the notions of disturbance (our signaling with communication) and fine-tuning (our "hidden" signaling). His corollary 3 states that "every classical model that reproduces the violation of a KS [Kochen-Specker] inequality in a no-disturbance phenomenon requires fine-tuning". This is the case in the example of Table 10.3.

Thanks to Jerome Busemeyer for directing our attention to Cavalcanti's work.

References

Aerts, D. (2014). Quantum theory and human perception of the macroworld. *Frontiers in Psychology, 5*, 1–19.

Aerts, S., Kitto, K., & Sitbon, L. (2011). Similarity metrics within a point of view. In D. Song, et al. (Eds.), *5th International Conference in Quantum Interaction* (pp. 13–24). Berlin: Springer.

Aerts, D., Gabora, L., & Sozzo, S. (2013). Concepts and their dynamics: A quantum-theoretic modeling of human thought. *Topics in Cognitive Science, 5*(4), 737–772.

Aspect, A., Dalibard J., & Roger, G. (1982). Experimental test of Bell's inequalities using time-varying analyzers. *Physical Review Letters, 49*, 1804–1807.

Atmanspacher, H., & Filk, T. (2006). Complexity and non-commutativity of learning operations on graphs. *BioSystems, 85*, 84–93.

Atmanspacher, H., & Filk, T. (2013a). The Necker-Zeno model for bistable perception. *Topics in Cognitive Science, 5*, 800–817.

Atmanspacher, H., & Filk, T. (2013b). Options for testing temporal Bell inequalities for menal systems. In D. Song, et al. (Eds.), *5th International Conference in Quantum Interaction* (pp. 128–137). Berlin: Springer.

Atmanspacher, H., & Römer, H. (2012). Order effects in sequential measurememts of non-commuting psychological observables. *Journal of Mathematical Psychology, 56*, 274–280.

Bell, J. (1966). On the problem of hidden variables in quantum mechanics. *Reviews of Modern Physics, 38*, 447–452.

Bruza, P. D., Kitto, K., Ramm, B. J., & Sitbon, L. (2015). A probabilistic framework for analysing the compositionality of conceptual combinations. *Journal of Mathematical Psychology, 67*, 26–38.

Carmi, A., & Cohen, E. (2018). Relativistic independence bounds nonlocality. arxiv.org/abs/1806.03607.

Cavalcanti, E. (2018). Classical causal models for Bell and Kochen-Specker inequality violations require fine-tuning. *Physical Review, 8*, 021018.

Cervantes, V. H., & Dzhafarov, E. N. (2018). Snow Queen is evil and beautiful: Experimental evidence for probabilistic contextuality in human choices. *Decision, 5*, 193–204.

Clauser, J. F., Horne, M. A., Shimony, A., & Holt, R. A. (1969). Proposed experiment to test local hidden-variable theories. *Physical Review Letters, 23*, 880–884.

Dzhafarov, E. N. (2003). Selective influence through conditional independence. *Psyhometrika, 68*, 7–25.

Dzhafarov, E. N., Kujala, J. V., Cervantes, V. H., Zhang, R., & Jones, M. (2016). On contextuality in behavioral data. *Philosophical Transactions of the Royal Society A, 304*, 20150234.

Filk, T. (2015). A mechanical model for a PR-Box. arXiv: quant-phys 1507.06789.

Filk, T. (2016). It is the theory which decides what we can observe (Einstein). In E. Dzhafarov et al. (Eds.), *Contextuality from physics to psychology* (pp. 77–92). Singapore: World Scientific.

Florig, M., & Summers, S.J. (1997). On the statistical independence of algebras of observables. *Journal of Mathematical Physics, 38*, 1318–1328.

Haven, E., & Khrennikov, A. Y. (2013). *Quantum social science*. Cambridge: Cambridge University Press.

Kochen, S., & Specker, E. P. (1967). The problem of hidden variables in quantum mechanics. *Journal of Mathematics and Mechanics, 17*, 59–87.

Kujala, J., & Dzhafarov, E. N. (2016). Probabilistic contextuality in EPR/Bohm-type systems with signaling allowed. In E. Dzhafarov, S. Jordan, R. Zhang, & V. Cervantes (Eds.), *Contextuality from physics to psychology* (pp. 287–308). Singapore: World Scientific.

Popescu, S., & Rohrlich, D. (2004). Nonlocality as an axiom. *Foundations of Physics, 24*, 379–385.

Pothos, E. M., & Busemeyer, J. R. (2013). Can quantum probability provide a new direction for cognitive modeling? *Behavioral and Brain Sciences, 36*, 255–274.

Pothos, E. M., Busemeyer, J. R., & Trueblood, J. S. (2013). A quantum geometric model of similarity. *Psychological Review, 120*, 679–696.

Tsirelson, B. (1980). Quantum generalizations of Bell's inequality. *Letters in Mathematical Physics, 4*(2), 93–100.

Walleczek, J., & Grössing, G. (2016). Nonlocal quantum information transfer without superluminal signalling and communication. *Foundations of Physics, 46*, 1208–1228.

Wang, Z., Solloway, T., Shiffrin, R. M., & Busemeyer, J. R. (2014). Context effects produced by question orders reveal quantum nature of human judgments. *Proceedings of the National Academy of Sciences of the USA, 111*, 9431–9436.

Chapter 11
Is There a Place for Consciousness in Quantum Mechanics?

Otávio Bueno

Abstract In this paper, I examine critically whether there is a role for consciousness in quantum theory. First, I consider von Neumann's (Mathematical foundations of quantum mech*anics* (The English translation, by Robert T. Beyer, of the original German edition was first published in 1955). Princeton University Press, Princeton, 1932) alleged introduction of consciousness in the interpretation of (non-relativistic) quantum mechanics, and conclude that consciousness plays no role in it. I then examine Wigner's (Symmetries and reflections: Scientific essays (Reprint from the 1967 edition published by Indiana University Press). Ox Bow Press, Woodbridge, 1979) views on the matter, identifying him, rather than von Neumann, as the leading proponent of the view that favors a prominent role for consciousness (see also Freire O, Jr: The quantum dissidents: Rebuilding the foundations of quantum mechanics (1950–1990). Springer, Dordrecht, 2015). I then question the aptness of such a role by advancing a minimalist interpretation of London and Bauer's (The theory of observation in quantum mechanics. In: Wheeler JA, Zurek WH (eds) Quantum theory and measurement. Princeton University Press, Princeton, pp 217–259. (The original work was published in French in 1939), 1939/1983) theory of observation in quantum mechanics, which has been taken as a key source of arguments in support for consciousness (particularly in connection with phenomenology, see French S: Stud Hist Philos Mod Phys 33:467–491, 2002), and conclude with a dilemma against views that identify a place for consciousness in quantum theory.

11.1 Introduction

Since John von Neumann's alleged introduction of consciousness in the interpretation of (non-relativistic) quantum mechanics (von Neumann 1932), the issue of the place of consciousness in the understanding of the theory has been alive and explored in various forms (see, for instance, London and Bauer 1939/1983; Wigner

---

O. Bueno (✉)
Department of Philosophy, University of Miami, Coral Gables, FL, USA

J. Acacio de Barros, C. Montemayor (eds.), *Quanta and Mind*, Synthese Library 414, https://doi.org/10.1007/978-3-030-21908-6_11

1979, pp. 171–184; Barrett 1999, pp. 47–55; French 2002; Freire 2015, pp. 142–155). In this paper, I consider whether there is a place for consciousness in quantum mechanics. In particular, I examine whether consciousness can be taken as a device to account for the collapse of the wave function, in the way that some take to have been defended by von Neumann (1932) (see Mould 1998). In what follows, I resist this interpretation, and suggest that this is a view that, suitably understood, stems from Wigner (1979), and *not*, as may be thought, from London and Bauer (1939/1983), at least on the minimalist interpretation I suggest of their work. I argue that, in light of the qualitative features of consciousness, it is unclear that any such role can be assigned to it, and that quantum mechanics does not leave much room for consciousness in the end.

11.2 Von Neumann on Measurement

In his discussion of measurement, von Neumann (1932/1955, p. 351) distinguishes two interventions that can occur in a system: (a) the evolution of a system over time (a determinist, continuous process described by Schrödinger equation), and (b) (what is often described as) the collapse of the state of the system upon measurement (an instantaneous, discontinuous process). Interestingly, von Neumann does *not* describe (b) as a "collapse". He simply calls it (1) and identifies it via its corresponding mathematical description. As is typical in his work, von Neumann avoids adding a gloss on the issues under consideration, thus resisting to provide an interpretation of them beyond what is strictly needed.

On von Neumann's view, every measurement ultimately requires an observer and it is something different from the environment in which it takes place. He notes that

> [M]easurement or the related process of the subjective perception is a new entity relative to the physical environment and is not reducible to the latter. (1932/1955, p. 418)

As understood by von Neumann, every measurement has a corresponding subjective trait (a "subjective perception") and is, thus, something distinct from the system (the "physical environment") that is being measured. The significance of this point is that if a measurement were not different from the system being measured, instead of being a measurement, it would be just part of the evolution of the system.

Von Neumann continues:

> [S]ubjective perception leads us into the intellectual inner life of the individual, which is extra-observational by its very nature (since it must be taken for granted by any conceivable observation or experiment). (1932/1955, p. 418)

We find here von Neumann's association between the subjective character of observation and the physical system under study—what he calls a "psycho-physical parallelism" (1932/1955, p. 419). On the one hand, the subjective perception is involved in any measurement, since it is required by any observation. On the other hand, this subjective perception is extra-physical, it is not part of the system under study. However, von Neumann emphasizes:

> [i]t must be possible so to describe the extra-physical process of the subjective perception as if it were in reality in the physical world—i.e., to assign to its parts equivalent physical processes in the objective environment, in ordinary space. (1932/1955, p. 419)

He suggests that we divide the world into three parts: the (physical) system, the measuring instrument, and the observer (von Neumann 1932/1955, p. 421). The line demarcating the observer from the rest can be drawn arbitrarily (but it needs to be drawn somewhere): the observer may include the measuring instrument and be separated from the physical system or the physical system and the measuring instrument can be placed together and be separated from the observer. This means that what counts as the observer is a moving target, provided that it always includes "the extra-physical process of the subjective perception" (1932/1955, p. 419).

It is important to note that von Neumann does *not* state that consciousness collapses the wave function: he does not talk of consciousness at all. He does not advance an interpretation of the issues beyond what is strictly required, and for which there is evidence. (In this respect, von Neumann is a good empiricist; see Bueno 2016.) Talk of "collapse" suggests that something *caused* such a collapse, and invites the question of what exactly is the *source* of the collapse in question. Consider the argument to the effect that what differentiates between the evolution of a quantum system and what goes on in a measurement is something that is not part of the system (otherwise it would not be a measurement of the system, but just part of its evolution, and the two processes would not be distinguished). In light of this, one is then led to conclude that, in a measurement, something of a different kind needs to be introduced. (Interestingly, even this argument is not explicitly formulated by von Neumann.) Although von Neumann does talk of the observer (but not of consciousness) and of an extra-physical process, he claims that one can treat the extra-physical process as though it was physical. As a result, his entire discussion is couched in strictly naturalistic terms.

Consciousness arguably has a particular phenomenology (it has specific qualitative features): there is something that is like to have such a consciousness. (As Thomas Nagel 1974 notes, it is consciousness that makes the mind-body problem intractable.) Consciousness is a subjective process, and attempts to describe it objectively miss the content of the phenomenon under consideration. In contrast, what von Neumann is considering is simply an extra-physical process, one that can be, in fact, described *as if it were in the physical world*. Clearly, consciousness does not fit that bill, and if something weaker than consciousness were invoked, one would be dealing with an item of a different kind altogether.

Before von Neumann developed his mathematical formulation of quantum mechanics, some physicists did describe the measurement problem in ways that assigned intentional, conscious traits to it and to the accompanying physical situation. In a discussion of the measurement problem, Paul Dirac (Institut Solvay 1928, p. 262; quoted in Barrett 1999, p. 25) noted that, according to quantum mechanics, the state of the world can be described by a wave function ψ that typically evolves deterministically, in the sense that the value it initially has determines all of its subsequent values. However, he noted, it is possible that, after a given moment, say, t_1, the wave function is decomposed into several wave functions that do not interfere with one another from that moment on. In this case, he continues, the wave function

ψ can be represented by a series of the form $\sum_n c_n \psi_n$, with no interference among the wave functions ψ_n. As a result, the relevant state of the world would not be properly described by ψ alone; rather this would be accomplished by one of the ψ_n. In Dirac's own words:

> One can say that nature chooses the particular ψ_n that is suitable, since the only information given by the theory is that the probability that anyone of the ψ_n will be selected is $|c_n|^2$. Once made, the choice is irrevocable and will affect the entire future state of the world. The value of n chosen by nature can be determined by experiment and *the results of all experiments* are numbers that describe such choices of nature. (Institut Solvay 1928, p. 262; quoted in Barrett 1999, p. 25)

The idea that nature "chooses" a particular wave function is a clear attribution of an intentional stance to the description of the relevant physical phenomenon. In just three sentences, the idea that nature is involved in a choice is referred to four times in a row. It is a salient trait of Dirac's account of the situation.

Dirac, of course, may not have been the only one to use an expression of this sort, and this expression can be taken only as an encoded portrayal of a more complex process in which intentional terms could be ultimately dispensed with. In this sense, the description should not be taken literally. But until such expressions are fully dispensed with—acknowledging the fact, noted by Dirac, that the relevant values can be determined experimentally and that, more generally, experimental results ultimately describe the numbers in question—it is the very idea that the numbers result from "choices of nature" that is problematic. This way of speaking identifies an agency (nature's choice processes) that do not seem to correspond to any feature of the world.

11.3 Wigner on Measurement

In contrast with von Neumann, Eugene Wigner does talk of consciousness and argues that not only do physical processes influence non-physical (conscious) ones, but that the same goes in the other direction too. As he notes:

> [T]he impression which one gains at an interaction, called also *the result of an observation*, modifies the wave function of the system. (Wigner 1979, p. 175)

And he continues:

> The modified wave function is, furthermore, in general unpredictable before the impression gained at the interaction has entered our consciousness: it is the entering of an impression into our consciousness which alters the wave function because it modifies our appraisal of the probabilities for different impressions which we expect to receive in the future. It is at this point that the consciousness enters the theory unavoidably and unalterably. (Wigner 1979, pp. 175–176)

Differently from von Neumann, Wigner clearly makes the point in terms of consciousness, and suggests that consciousness "alters the wave function". But how is this possible?

It seems that in order to run the line that Wigner intends a form of idealism needs to be brought to the forefront (dualism seems to be involved as well). As he points out:

> There are several reasons for the return, on the part of most physical scientists, to the spirit of Descartes's "*Cogito ergo sum*", which recognizes the thought, that is, the mind, as primary. (Wigner 1979, p. 172)

And more explicitly, he notes:

> When the province of physical theory was extended to encompass microscopic phenomena, through the creation of quantum mechanics, the concept of consciousness came to the fore again: it was not possible to formulate the laws of quantum mechanics in a fully consistent way without reference to the consciousness (Wigner 1979, p. 172)

However, as argued above, strictly speaking, it is not consciousness that is required in this context, at least not as far as von Neumann's formulation of quantum theory is concerned, given that only the observer is involved, which is then treated as an extra-physical process that is described as though it were in the physical world. Wigner does quote von Neumann regarding the malleability of the divide between observer and physical system (see Wigner 1979, p. 172, note 4). But the general point about consciousness does not go through.

Interestingly, Wigner also invokes a passage from Werner Heisenberg in support of the indispensability of consciousness in the formulation of quantum mechanics. According to Heisenberg:

> The laws of nature which we formulate mathematically in quantum theory deal no longer with the particles themselves but with our knowledge of the elementary particles. (Heisenberg 1958; cited by Wigner 1979, p. 172, note 3)

Wigner highlights that the "'our' in this sentence refers to the observer who plays a singular role in the epistemology of quantum mechanics" (Wigner 1979, p. 172, note 3). But, as we saw in the discussion of von Neumann's view, the introduction of the observer and the introduction of consciousness are strictly different, since the observer can be understood as part of the physical world under study. Furthermore, Heisenberg's remark is not particularly about consciousness: he is acknowledging the role of the observer and making an epistemological point to the effect that one's knowledge of elementary particles ultimately depends on the observer. Even if it is granted that knowledge of such particles depends on the observer (most pieces of knowledge do, in some way or another), it does *not* follow, however, that consciousness and its qualitative features are required for that.

11.4 London and Bauer on Measurement

It may be argued that Wigner is here following the lead from Fritz London and Edmond Bauer in their work on the theory of observation in quantum mechanics, in which they seem to identify a clear role for consciousness in quantum theory (London and Bauer 1939/1983). In fact, in an insightful essay, Steven French argues, very compellingly, that one needs to read London and Bauer's work in terms of

Husserl's phenomenology (French 2002). According to French, we ought to take seriously that when London and Bauer (particularly London, who is well known for highlighting the philosophical underpinnings of his work) refer to an "I" (an ego) in their essay, when they use expressions such as "immanent knowledge" (and, following French, I will quote the relevant passages below), they are explicitly employing a vocabulary that is best understood in terms of their phenomenological roots.

I find French's reading extremely reasonable and perceptive, but I would like to suggest a different interpretation, one that is a bit more minimalist and which does not project such a heavy and philosophically contentious framework (Husserl's phenomenology) to the London and Bauer essay. Despite London's explicit philosophical preferences, Bauer, as French himself acknowledges (2002, p. 473, note 14), does not seem to have had much interest in philosophical issues. I suspect that Bauer probably would be more sympathetic to a philosophically minimalist account of his work with London than one that adds substantial philosophical assumptions to the approach. Now, admittedly, a crucial section of London and Bauer's (1939/1983) article, which focuses on the act of objectivation, as French notes (in private correspondence), appears to have been written by London. But given that both physicists are identified as authors of the paper, it is important to offer an account that provides a common ground to both. Since I intend to advance an alternative reading to the one developed by French, in what follows I will accompany very closely French's own portrayal of London and Bauer's article and his rich discussion of it, even though I end up offering a different, less philosophically committed, interpretation of the work.

London and Bauer start by examining the measurement of a given quantity F of a quantum system which is in the state $\psi = \sum_k \psi_k u_k(x)$, with u_k as an eigenfunction that corresponds to the value f_k of the quantity F that is being measured (London and Bauer 1939/1983, p. 250; French 2002, pp. 482–483). To measure F, one needs a suitable apparatus (capable of measuring that quantity) so that, after an interaction between the apparatus and the system, the apparatus' pointer has its eingenvalues in a one-to-one relation with the system's values f_k. Clearly, this is the outcome of the interaction between the system and the apparatus. However, to be a measurement, rather than a simple interaction, something more is needed. A measurement, London and Bauer note, "is achieved only when the position of the pointer has been *observed*" (London and Bauer 1939/1983, p. 251; French 2002, p. 483). This means that the *observation* of the recorded result is ultimately required for a measurement.

London and Bauer then elaborate:

> It is precisely this increase of knowledge, acquired by observation, that gives the observer the right to choose among the different components of the mixture [that is, in today's terminology, the superposition; see French [2002], p. 483, note 27] predicted by theory, to reject those which are not observed, and to attribute thenceforth to the object a new wave function, that of the pure case which he has found. (London and Bauer [1939/1983], p. 251; quoted in French [2002], p. 483)

Curiously, London and Bauer note that the observer has "the right to choose" among the various components of a superposition. One may think that this is a place in which intentionality and perhaps even consciousness will play a role. Does this mean that it is the observer who chooses which component of a superposition obtains? The answer is clearly negative. The system and the measuring apparatus interact; a measurement is made; the result is observed; the observer rejects those values that are not detected and adjust the description of the system to the new wave function corresponding to the measurement that was made. The choice made by the observer consists in assigning to the system the result of that measurement. This is made *after* the measurement is obtained, and it is by no means what prompted the result of the measurement in the first place. The choice involved is the choice to maintain the coherence of the measurement results, by assigning to the system that is measured the observed results of the measurement. No revisionism regarding a fundamental role for consciousness is needed for that.

London and Bauer do see here, however, "the essential role played by the consciousness of the observer in this transition from the mixture to the pure case" (1939/1983, p. 251; quoted in French 2002, p. 483). It is in this context, in particular, that French emphasizes the central role played by phenomenology in shaping the understanding of London and Bauer's work. Let us examined how this issue unfolds.

To the ensemble formed by the combination of the object x (the quantum system under study), the measuring apparatus y, and the observer z, there is a corresponding global wave function $\Psi\ (x, y, z) = \sum \psi_k u_k(x) v_k(y) w_k(z)$, with u_k, v_k, and w_k representing, respectively, the different states of the system, the apparatus, and the observer (London and Bauer 1939/1983, p. 251; French 2002, p. 483). London and Bauer note:

> Objectively—that is, *for us* who consider as 'object' the combined system x, y, z—the situation seems little changed to what we just met when we were considering only apparatus and object. (London and Bauer [1939/1983], p. 251; quoted in French [2002], p. 483)

This is a suggestive remark since it indicates an objective separation between the quantum system that includes the observer from the one that just has the object and the apparatus. The passage also assumes that London and Bauer (the "us" that are referred to in the text) are not themselves the observers in question. This allows them to treat the system that includes the observer as not being essentially different from the one that does not.

However, from the perspective of the *observer*, the situation is very different. After all, it is the observer who is interacting with the system, collecting the information from the measurement, and drawing the relevant inferences from what is observed. As London and Bauer point out:

> The observer has a completely different impression. For him it is only the object x and the apparatus y that belong to the external world, to what he calls "objectivity". By contrast he has with himself relations of a very special character. He possesses a characteristic and quite familiar faculty which we can call the "faculty of introspection". He can keep track from moment to moment of his own state. By virtue of this "immanent knowledge" he attributes to himself the right to create his own objectivity—that is, to cut the chain of statistical correlations summarized in $\Psi\ (x, y, z) = \sum \psi_k u_k(x) v_k(y) w_k(z)$ by declaring "I am in the state w_k" or more simply, "I see $G = g_k$" or even directly, "$F = f_k$" (London and Bauer [1939/1983], p. 252, quoted in French [2002], p. 483).

In this passage, London and Bauer identify the special role the observer plays in the measurement process. As noted, French interprets the reference to "immanent knowledge" as something that ought to be read phenomenologically. But there is also a more deflationary and philosophically less controversial interpretation. It goes as follows.

Observers have privilege access to their own mental states: they can introspect and, in this way, be aware of what state they are in. This is expressed by London and Bauer's reference to a "faculty of introspection". (Of course, one may add, introspection is not infallible, and it is possible for one to misinterpret one's own mental states. I may think, for instance, that I am indifferent to the outcome of a soccer game and be surprised to discover how pleased I was with the final result. This suggests that indifference was not the proper state I was actually in all along; see Williamson [2000].) With the awareness of one's mental states (the "immanent knowledge" London and Bauer refer to), the observer is in a position to "create [one's] own objectivity". That is, one is licensed to move through a series of inferences from the observer's own mental state through the state of the measuring apparatus all the way to the state of the physical system being measured. The result of this process is the observer's evidence (the observer's "objectivity").

This is accomplished by cutting the "chain of statistical correlations" provided by the relevant wave function of the ensemble of object, apparatus, and observer. (The "statistical correlations" in question are understood as the correlations between the various states of the observer, the apparatus, and the object.) The observer's declaration "I am in the state w_k" is precisely the awareness of the state one is in upon observing the outcome of the measurement process. That outcome is expressed by the particular position of the apparatus' pointer G and the corresponding eigenvalues g_k. This leads the observer then to note: "I see $G = g_k$". From this, the observer infers that the state of the quantum system is given by the corresponding identity between the system's state F and the suitable eigenvalues f_k, thus obtaining the identity "$F = f_k$". In this way, the observer moves from one's own state to the apparatus' state all the way to the state of the system being measured.

This is a series of inferential steps that are ultimately grounded in the statistical correlations between each of the various states. It is a process of making an inner state objective in virtue of the correlations between the inner state and those that are not, namely, the states of the apparatus and of the system being measured. And this entire process is prompted by the new piece of information that the observer obtained as the result of the measurement and which was provided by the apparatus that interacted with the quantum system in question.

London made this point explicit in a note he added to his own copy of the essay with Bauer:

> Accordingly, we will label this creative action as "making objective". By it the observer establishes his own framework of objectivity and acquires a new piece of information about the object in question. (London and Bauer [1939/1983], p. 252, quoted in French [2002], p. 483)

What is at issue, then, is the way in which the observer uses the information that is obtained through the measurement process to infer the state in which the physical system is in. The observer "establishes [one's] own framework of objectivity" in the sense that the information about the measurement outcome that is obtained is used to account for the objective state of the physical system. But that is a choice of the observer, who needs to draw the relevant inference rather than reject (or ignore or alter) the result of the measurement.

Interestingly, London and Bauer highlight precisely this point in a passage that French also emphasizes:

> [...] it is not a mysterious interaction between the apparatus and the object that produces a new ψ for the system during the measurement. It is only the consciousness of an "I" who can separate himself from the former function Ψ (x, y, z) and, by virtue of his observation, *set up a new objectivity* in attributing to the object henceforward a new function $\psi(x) = u_k(x)$. (London and Bauer [1939/1983]; emphasis in the original; quoted in French [2002], pp. 483–484)

Explicitly rejected here is the idea that there is any "mysterious interaction" between the object being measured and the measuring apparatus that somehow creates a new wave function for the resulting system. There *would be* such a mysterious interaction if consciousness were indeed responsible for the production of such a wave function. After all, how could something arguably non-physical have such an effect? (Of course, for the phenomenologist, the relation between consciousness and the world is not mysterious either, although what is taken to be the *world* is significantly reconceptualized within phenomenology.) However, rather than taking consciousness to have any such a causal role, London and Bauer emphasize that the observer, by being aware of the state one's in, is then able to separate oneself from the ensemble of object, apparatus, and observer (something that is crucial to the phenomenological interpretation), in order then to infer, from the information obtained in the measurement process, what the wave function of the object being measured ultimately is (taken to be), namely, $\psi(x) = u_k(x)$. The observer is then able to "set a new objectivity" by externalizing, from the observation of the interaction between object and apparatus, the (inferred) wave function of the object being measured.

Note that the only role that consciousness plays here does not seem to be particularly tied to quantum mechanics. It is simply the awareness of the state the observer is in that is needed to initiate the process of attributing to the object being measured the result of the measurement process. This is a way of ensuring the coherence of the observation obtained through measurement and the corresponding wave function of the object being measured. To achieve that none of the qualitative traits of consciousness are needed at all. Rather, articulated here is what turns out to be primarily a methodological maneuver (to ensure coherence).

It may be objected that, on this reading of London and Bauer's work, there is nothing about quantum mechanics that makes the observer particularly special, and that virtually the same account could be offered in the case of the measurement process in classical physics, for instance. In response, there is indeed an important generality in London and Bauer's approach to measurement that could be easily transferred to measurements in the context of other physical theories. But there is still something special about quantum mechanics, given the delicate nature of quantum systems and how extraordinarily sensitive they are to interference. This requires, in particular, special attention by the observer when drawing inferences about the state of the system from the information obtained in the measurement process itself. And this is precisely the point that London and Bauer correctly highlight. Once again, it is not consciousness, in any significant sense, that is relevant to the outcome in question.

It may also be complained that the main problem with the minimalist account is that it leaves entirely open how the observer can get, in the first place, a definite result after a measurement of a quantum system (this eventually allows the observer to become aware of the result in question). In other words, the minimalist account offers no solution to the measurement problem in quantum mechanics. In contrast, the phenomenological reading provides such a solution (for details, see French 2002). (I owe this objection to Steven French.)

In response, the minimalist account does not attempt to solve the measurement problem. It should not be part of an account of observation in quantum mechanics that it solves (or dissolves) such a problem: these are two separate issues. The former—about a theory of observation—concerns the role of observation and of the observer in quantum theory (and to address this issue is arguably the main goal of London and Bauer's 1939/1983 paper); the latter—about the measurement problem—considers how to reconcile two incompatible evolutions of quantum systems. The minimalist account identifies what is involved in acts of observation in quantum mechanics and the role played by the observer in them in a way that is neutral regarding possible solutions to the measurement problem. This highlights an additional sense of minimalism involved here: nothing more than what is strictly needed to make sense of observation is advanced by the proposal. To go beyond this and attempt to solve the measurement problem would involve additional, and significantly controversial, assumptions that are not at all required to account for the role played by the observer in quantum mechanics.

11.5 Conclusion

In discussions of the role of consciousness in quantum mechanics, the qualitative aspect of consciousness is often not taken seriously. But without engaging carefully with this aspect, it is unclear that what one is considering is really consciousness. Even though observers have consciousness, it is not their consciousness that is doing the relevant work, not at least in the formulation of quantum mechanics advanced by von Neumann or even London and Bauer. Wigner seems to want more, but it is unclear that anything more than just an extra-physical component that would be treated as though it was part of the physical world (as in von Neumann's formulation) is, in fact, needed.

In conclusion, a dilemma is reached: Either consciousness is taken seriously in the foundations of quantum mechanics or it is not. If it is taken seriously, then given its qualitative features, consciousness *cannot* play a role in quantum mechanics (since this would undermine the objective traits of the relevant descriptions). If consciousness is *not* taken seriously, then it is not consciousness but something else that plays a role in quantum mechanics (such an observer treated as part of the physical world). In either case, there does not seem to be much room for consciousness in the foundations of quantum theory.

Acknowledgements My thanks go to Acacio de Barros, Carlos Montemayor, and especially Steven French for extremely helpful discussions and correspondence on the issues examined in this work. French also sent me insightful comments on an earlier version of this article, which led to significant improvements. My thanks are also due to the audience at the conference "Quanta and Consciousness" that was held in San Francisco in April 10-11, 2018. Their comments and suggestions were invaluable.

References

Barrett, J. (1999). *The quantum mechanics of minds and worlds*. Oxford: Oxford University Press.

Bueno, O. (2016). Von Neumann, empiricism, and the foundations of quantum mechanics. In D. Aerts, C. de Ronde, H. Freytes, & R. Giuntini (Eds.), *Probing the meaning and structure of quantum mechanics: Superpositions, semantics, dynamics and identity* (pp. 192–230). Singapore: World Scientific.

Heisenberg, W. (1958). *Physics and philosophy: The revolution in modern science*. London: Penguin.

Freire, O., Jr. (2015). *The quantum dissidents: Rebuilding the foundations of quantum mechanics (1950–1990)*. Dordrecht: Springer.

French, S. (2002). A phenomenological solution to the measurement problem? Husserl and the foundations of quantum mechanics. *Studies in History and Philosophy of Modern Physics, 33*, 467–491.

Institut Solvay, Conseil de Physique. (1928). *Électrons et photons: Rapports et discussions du cinquième Conseil de physique tenu à Bruxelles du 24 au 29 octobre 1927 sous les auspices de l'Institut International de Physique Solvay*. Paris: Gauthier-Villars.

London F, Bauer E (1939/1983) The theory of observation in quantum mechanics J.A. Wheeler and W.H. Zurek (eds.), Quantum theory and measurement (Princeton: Princeton University Press), 217–259. (The original work was published in French in 1939)

Mould, R. (1998). Consciousness and quantum mechanics. *Foundations of Physics, 28*, 1703–1718.

Nagel, T. (1974). What is it like to be a bat. *Philosophical Review, 83*, 435–450.

von Neumann, J. (1932). *Mathematical foundations of quantum mechanics* (The English translation, by Robert T. Beyer, of the original German edition was first published in 1955). Princeton: Princeton University Press.

Wigner, E. (1979). *Symmetries and reflections: Scientific essays* (Reprint from the 1967 edition published by Indiana University Press). Woodbridge: Ox Bow Press.

Williamson, T. (2000). *Knowledge and its limits*. Oxford: Oxford University Press.

Chapter 12
Quantum Mechanics and Consciousness: Some Views from a Novice

Emmanuel Haven

Abstract In this purely conjectural paper we attempt to define how consciousness may be related to a measure of information. We use the quantum-like approach to formulate a possible relationship.

12.1 Introduction

The game 'Go' is probably amongst the most complex games humans can play. In 2017, the M.I.T. Technology Review (Condliffe 2017) reported that the AI company DeepMind had created a programme named 'AlphaGo Zero'. A very noteworthy fact about this programme is that it can develop playing strategies[1] in 'Go', *without* any input from human beings. An even more interesting fact is that this same programme can be successfully applied to other games. We note that this is a construct from AI. There is no consciousness involved. Or is there? How does the playing of strategies outsmarting humans in 'Go' compare to playing a beautiful piano concert of Scriabin? Can the computer play like Arthur Rubinstein? Steven Pinker remarks (p. 1) "... brains have the benefit of a billion-year research-and-development effort in which evolution equipped them with cheat sheets for figuring out how to outmaneuver objects, plants, animals and other humans." However, 'AlphaGo Zero' is built on the basis of no cheat sheets at all. No human input, at all, is needed for the computer to outsmart humans in a very complex game indeed.

There may be scope to think about imputing levels of consciousness into machines. Dehaene et al. (2017) discuss two levels of conscious computation, C1 and C2. C1 refers to the idea of availability. In the words of the authors this is akin

[1] AlphaGo is an AI programme which beats human players at 'Go'.

E. Haven (✉)
Memorial University, St. John's, NL, Canada
IQSCS, Leicester, UK
e-mail: ehaven@mun.ca

J. Acacio de Barros, C. Montemayor (eds.), *Quanta and Mind*, Synthese Library 414, https://doi.org/10.1007/978-3-030-21908-6_12

(p. 486) to "having the information in mind." The authors call C2, self monitoring (p. 487): "the cognitive system... obtain(s) information about itself.". As the authors remark, computations may neither need C1 nor C2. However, imputing C1 and C2 into a machine would not be all that difficult. They give the example of a low gas light in a car. C1 would be synonymous with the low gas light communicating with say a GPS system to find gas stations which are near. C2, in the words of the authors would then, in the context of this example (p. 491), "...keep a list of its subprograms, compute estimates of their probability of succeeding at various tasks...".

From the outset, let us ask a superhard question: 'what causes consciousness'? Is the brain the main tool by which consciousness gets awakened? Or not? Donald Hoffman (2005) remarks: "...there is as yet no physicalist theory of consciousness, no theory that explains how mindless matter or energy or fields could be, or cause, conscious experience." From an intuitive point of view, one would sense that the brain is the key ingredient to consciousness. Donald Hoffman further remarks "Yet neither the brain nor the NCC (Neural Correlates of Consciousness) causes consciousness. Instead consciousness constructs the brain and the NCC".

In the next section of the paper, we briefly discuss NCC's. In the section following we look at the centrality of the concept of information in consciousness. We then discuss how integrated information and active information may be connected with each other. The section after that provides for some ideas on how consciousness may be seen in the ontological interpretation of quantum mechanics. We then continue with the idea of Fisher information as a measure of reduction of uncertainty and lay out, in the next section, its link with the quantum potential and wave function. We then discuss in the section following, how time asymmetry appears within the context of consciousness and we then conclude.

12.2 What Are NCC's?

What are NCC's? There are many sources defining NCC's. We use Tononi and Koch (2008, p. 246), where they quote Crick and Koch (1995a, 1998b, 2003): "...neural correlates of consciousness (NCC) – the minimal neuronal mechanisms that are jointly sufficient for any one specific conscious percept." And they add that (p. 247) "every phenomenal, subjective state will have associated NCC."

Tononi and Koch clearly indicate that consciousness relates to the brain (p. 240) "...without requiring any obligatory interaction with the environment or the body." The authors identify levels of consciousness and we are all well aware (at least intuitively) of some of those levels. In sleep and especially in the measurement of reactions to the administration of anesthetics one can get a good sense of levels of unconsciousness. The authors report for instance, that there are various degrees of diminution of consciousness which can be related to levels of so called minimum alveolar concentration (MAC).[2] What is also interesting to note is their reporting of a fairly abrupt change from levels of consciousness to a loss of consciousness. This loss of consciousness can be triggered by what Tononi and Koch (2008) call a "deactivation of the cortex" (p. 244). However, they also note that (p. 248), "changes in neural activity do not necessarily correlate with changes in conscious experience."

12.3 The Centrality of Information When Discussing Consciousness

Where the idea of consciousness starts getting closer to the main argument we are seeking, is where Tononi and Koch claim that (p. 253) "the most important property of consciousness is that it is extraordinarily informative. This is because, whenever you experience a particular conscious state, it rules out a huge number of alternative experiences. Classically, the reduction of uncertainty among a number of alternatives constitutes information." It is precisely here where we start connecting with the objective of this paper. In Tononi (2004, p. 2) effective information is precisely defined and it refers to a measurement of information relative to a so called 'integrated' information structure. As an example, if I see an object and I perceive it as a 'white coloured pencil' then the information generated by the reduction of uncertainty (i.e. identifying the object as a white pencil), is called integrated if "it is above and beyond the information that is generated independently within the parts themselves." (Tononi and Koch 2008, p. 254). Here in our example, the information pertaining to the 'parts' is characterized by the information associated with (i) a pencil and (ii) the colour white, separately. Integrated information is a measure of consciousness. As reported in Tononi (p. 253): "...information associated with the occurrence of a conscious state is integrated information." This leaves completely aside the question of the kind of consciousness which is addressed as the second problem in Tononi (2004, p. 6).

Let us sum up: (i) a conscious state is akin to a reduction of uncertainty (but a quite peculiar one); (ii) the information associated with the conscious state is integrated information; (iii) quantity of consciousness is determined by quantity of information.

---

[2]Appleby (2014) mentions another scale, the Glasgow Coma Scale which (p. 13) "is widely used to quantify the level of consciousness in cases of brain damage." (see also Teasdale and Jennett (1974, 1976) as quoted in Appleby).

12.4 How Integrated Information and Active Information May Be Connected

Integrated information makes us think about a concept which saw its birth in physics, i.e. the idea of 'active information'. Let us attempt to explain this a little more. Active information is a key concept in the so called ontological interpretation of quantum mechanics. As mentioned in Pylkkänen (2015), active information relates a quantum field (its shape) to an electron. The field 'informs' the energy of the particle. The key references for this interpretation of quantum mechanics are Bohm (1952a,b), and Bohm and Hiley (1993).

At the macroscopic scale an analogy to active information could be interpreted as follows. We cite an example from Bohm (1990, p. 281) (see also Pylkkänen 2015). A dangerous criminal is roaming in a neighborhood. One sees Pylkkänen (2015, p. 329)[3]: "a suspicious-looking shadow. If your brain-mind interprets this as the 'assailant', meaning 'danger!', a powerful physico-chemical activity is likely to start in the brain and the body." This is an example of active information at the macroscopic level: information directs much higher energy (i.e. run away). This example can be connected to the discussion we had on integrated information and consciousness. Integrated information is generated here from the causal interaction between the observed shadow and the (imaginary) presence of the assailant. We may argue that the reduction of uncertainty is larger when the causal link is established between the two components, than if we were to look at the uncertainty reduction from each component individually.[4] In this example, consciousness is the prerequisite for the effect of active information to take place: i.e. effectuate the act of running, after the individual has determined integrated information (associated with his/her conscious state). We note that making consciousness a prerequisite in the setting we describe, may well be at odds with research in the area of consciousness. Pylkkänen (2017, p. 294) mentions that van Gulick (2014), Sect. 6.1 indicates that in work by Velmans (1991) consciousness "is neither necessary for any type of mental ability nor does it occur early enough to act as a cause of the acts or processes typically thought to be its effects."

Pylkkänen (2015, p. 329) also indicates that Bohm had in mind that human beings could be described by levels (from what he calls manifest (where the physical is more dominant) to subtle (where the mental is more dominant)). This could also be related to the second problem Tononi (2004, p. 6) identified in his work: the kind of consciousness. The modality and submodality are key concepts here. Hearing is an example of modality, whilst hearing low frequencies is an example of a submodality. The quality of consciousness lays out ranges on the submodalities and as Tononi (2004, p. 7) also mentions learning affects the quality of consciousness.

[3] See Bohm (1990, p. 281), where that example is mentioned.

[4] It can be debated whether the claim we make is true. The assailant himself, as the source of the shadow, may contain more information than the shadow coming from the assailant's presence and the light falling over his physical presence.

As an example, if one has never played the piano, the very first time one plays gives a different quality of consciousness (submodality is affected) than compared to the case where we play the instrument for the twenty second time. Integrated information seems not only to determine the quantity of consciousness it also pinpoints to the quality of consciousness. Can the modalities of consciousness relate to Bohm's idea of the manifest level (i.e. physical pain for instance) towards the subtle (i.e. the seeing of color)?[5]

12.5 Consciousness Defined in the Ontological Interpretation of Quantum Mechanics

How is consciousness defined in the Bohmian framework (i.e. the ontological interpretation of quantum mechanics)? Pylkkänen (2015, p. 332) asks an important question: "...how could such a 'very subtle' field carrying information possibly be able to act upon the more manifest processes, e.g. in the motor cortex?".[6] Here we need to quote (Bohm 1990, p. 283) "...that which we experience as mind, in its movement through various levels of subtlety, will, in a natural way ultimately move the body by reaching to the level of the quantum potential and of the 'dance' of the particles. There is no unbridgeable gap or barrier between any of these levels. Rather at each stage some kind of information is the bridge." This is an important quote as it does mention the quantum potential to which we come back later in this paper. Bohm (1990, p. 283) also mentions: "The content of our consciousness is then some part of this over-all process".

Let us come back to the idea of 'active information'. In Bohm (1990), we can read (p. 281): "...the whole notion of active information suggests a rudimentary mind-like behaviour of matter, for an essential quality of mind is just the activity of form, rather than substance." He juxtaposes this statement with what he calls "mechanical kind of interaction" of independent parts in classical physics. We can sense, although Bohm does not define consciousness really explicitly, how consciousness is definitely farther removed from classical physics. The essence of active information may well be contained in this quote (Bohm 1990, p. 282): "...in the context of the processes of thought, there is a kind of active information that is simultaneously physical and mental in nature. Active information can thus serve as a kind of link or 'bridge' between these two sides of reality as a whole. Those two sides are inseparable in the sense that information contained in thought, which we feel to be on the 'mental' side, is at the same time a related neurophysiological, chemical and physical activity..." This quote raises important issues. The link between mental and physical is of course of key importance. In the work of de

[5]At the level of submodalities, there is debate on where to locate the experience of a submodality, like when one sees a color. See Skokowski (2004).

[6]The very subtle field refers to some analogy of a quantum field.

Barros and Oas (2017), the authors discuss whether the interaction between mind and matter is the source of the collapse of the wave function. In Atmanspacher (2012) mention is made of this relationship within the explicit context of the exchange of ideas between two luminaries of, on the one side, psychology (C. Jung) and on the other side, physics (W. Pauli). This led to a discussion of the mind-matter relationship using the physics concept of complementarity. See also Khrennikov and Haven (2013) and Haven and Khrennikov (2013).

12.6 Fisher Information as Reduction of Uncertainty

In Sect. 12.3, we mentioned that quantity of consciousness could be measured by integrated information. A conscious state is akin to a reduction of uncertainty. One measure of information which precisely measures the level of noise, is Fisher information. It is a very intuitive measure of information which, in essence, gives a lot of weight on how a probability density function behaves over several measures of the noise component when the observed value is decomposed in the true value and noise. The key ingredient for the measurement of this type of information is the slope of the probability density function (PDF) function over the noise parameter. Let us briefly consider a very extreme case, the Dirac-Delta function[7] as PDF. In that case, the slope of the PDF is extremely steep and the Fisher information measure will be very high. This is intuitive, since we are in effect saying there is almost no fluctuation on noise and there is thus a lot of information. In the other extreme case, if the PDF is almost horizontal over the noise parameter, we mean we know virtually nothing about the noise parameter and it can be 'anything'. There are huge fluctuations. The level of information, I, thus relates to a measure of error (err) and it can be defined as the (squared) distance between the estimate of the true value and the true value. In social science, the so called Cramer-Rao inequality is well known and it says that $err^2 I \geq 1$: a trade off between large error and little information (and vice versa). Is it now presumptuous to consider this specific measure of information (amongst the many measures of information available), as a reasonable candidate for the measurement of consciousness? If we read onto the next section, the rationale for using I is, we think, provided by the link of this measure with the quantum potential we started discussing in Sect. 12.5.

12.7 The Link with the Quantum Potential of the Ontological Interpretation

In Sect. 12.5, we introduced the notion of quantum potential. From the outset, we need to heed the cautionary words of Bohm (1990), when he discusses how

[7]The Dirac Delta function, although called 'function' it is not a function in a mathematical sense.

to possibly relate mental processes to quantum mechanics. Says Bohm (1990, p. 283) "To bring this about, one could begin by supposing,... that as the quantum potential constitutes active information that can give form to the movements of particles, so there is a superquantum potential that can give form to the unfoldment and development of this first order quantum potential." Bohm then very explicitly indicates that this superquantum potential would not work within quantum theory. And quantum theory would be an approximation if "the action of the superquantum potential can be neglected." (ibid.).

Fisher information as defined in Sect. 12.6, can be shown to be narrowly linked to the quantum potential. A key paper by Reginatto (1998) shows the relationship. The quantum potential is defined as (omitting the Planck constant): $\frac{1}{R}\nabla^2 R$, where R is the amplitude of the so called polar form of the wave function. In the consciousness setting we are in, heeding Bohm's remark, the wave function is very probably not of a quantum mechanical flavour. In fact, within the Bohmian framework, Stapp (1996) makes the important remark that (p. 198) "in Bohm's theory the contents of our consciousness is determined by what the particle part of the universe is doing, not the wave part."

If the wave function were of importance, we then enter a debate of what the wave function represents. This is a difficult and long-ongoing exchange of ideas which puts one right into the foundations of quantum mechanics. Goldstein (2010) asks poignant questions about the wave function such as whether it is subjective or objective and especially, as Goldstein (2010, p. 336) queries: "if it is objective, does it represent a concrete material sort of reality, or does it somehow have an entirely different and perhaps novel nature?" We have in our work (Haven et al. 2017) proposed a quantum-like paradigm where essentially we accept the wave function as being information and the quantum mechanical formalism can be seen as a formalism by which we can show a mechanics for the processing of information.

12.8 Time Asymmetry and Consciousness

Nelson (1966, 2014) provides for an approach which is sometimes known as the stochastic interpretation of quantum mechanics. In his work he defines a drift function (as part of a Brownian motion[8]) as either defined (i) as an expectation of what we could call a difference between a future value and a present value (of say a position) (forward time expectation) or (ii) an expectation of what we could call a difference between a present value and a past value (of say a position) (backward time expectation). He then defines two velocity fields: (i) a mean velocity field (an average of both expectations defined above); and (ii) a so called osmotic velocity

[8]A Brownian motion can be seen as a stochastic differential equation composed of a drift and a diffusion. It describes a time dependent stochastic process.

field; which is the difference of both expectations as defined above.[9] The key issue is now as follows. The forward time and backward time expectations as defined above are exactly equal to each other in Newtonian mechanics. But when we step out of Newtonian mechanics, the difference is set as non-zero and the osmotic velocity is non-zero. This indicates an asymmetry in time: i.e. an asymmetry between past and future time. This is not uninteresting for our purposes. If we think of the degrees of consciousness which for instance, within the setting of anesthetics (as discussed in Sect. 12.2) relate to levels of minimum alveolar concentration (MAC), then our degrees of consciousness do also impute different time scales. The asymmetry in forward and backward time is quite striking and intuitive. What is also striking is that this difference between forward and backward time is narrowly related to Fisher information. Without this asymmetry between past and future we can not make this link to Fisher information.

So we can sum up:

- asymmetry between forward and backward time seems a needed ingredient (at least intuitively) when one thinks of levels of consciousness and the imputed time scales corresponding to those levels
- this asymmetry is also linked to the existence of Fisher information
- Fisher information is intimately related to the quantum potential

The above bullet points are giving a conjectural story. There is no proof. But if we relate the bullet points to the arguments that Bohm made in relation to consciousness, there are maybe some ingredients for the beginning of a formalism for the measurement of levels of consciousness within a wider setting of information.

It is key, I think, to remind oneself of the so called 'RWR principle' (Reality without Realism) that Andrei Khrennikov and Arkady Plotnitsky developed (2015). We can probably not in the mind-matter problem, account for a mechanical description of how mind and matter connect. Such absence of mechanical description calls in the RWR principle. The background of the principle is maybe best described as: "The existence or reality of quantum objects, a form of reality beyond representation or even conception, is inferred from effects they have on our world..." Plotnitsky (2017, p. 4). Maybe this statement ties in also to the hard problem of consciousness. Chalmers's provided for a distinction between the 'easy' and 'hard' problems. We discussed above submodalities like seeing red, but what does it mean to, in the words of Chalmers (2010, p. 5) "experience visual sensations: the felt quality of redness..." In Chalmers' words: "The really hard problem of consciousness is the problem of experience." (Ibid.) Shall RWR occur at the level of the hard problem? This is a question one might pose. We conclude with some words out of Appleby (2014, p. 35): "Discussions of qualia are often vitiated by the idea that there are two pictures involved: one that is colored (the picture we get from our eyes) and one that is not (the picture we get from physics).....neither of these pictures exist....the mathematical descriptions which physics gives us are not pictures either" Does this

[9]This difference is divided by 2.

go back to the RWR idea: a quantum object's existence is inferred from effects it has on us (or the world)? No. The RWR approach still does believe there is a mathematical description. What (Appleby 2014, p. 36) questions is the identification "of reality with the mathematical description" My understanding is that this applies to the venture of wanting to explain consciousness along that route. I can not vouch for either opinion whether this very hard problem of defining consciousness is definable – mathematically or not.

However, in recent work Andrei Khrennikov (2015) considers the modelling of the *unconscious* inference from a sensation to form a perception. It is quite remarkable to see how the use of a tensor product leads naturally to define a mental state and how the idea of entanglement between perception and sensation can be formalized. The construction of a positive operator valued measure as a formal representation of such inference, shows that formalisms can be used to positive effect.

12.9 Conclusion

This paper has laid out some ideas on how consciousness and elementary ideas from quantum mechanics can be connected. This is not in any way a new attempt. The literature is replete with very serious attempts to define consciousness. Let us finish this paper with a quote from Stapp (1996, p. 210) which implicitly shows the challenge we are facing: "The complexity of a human experience is a consequence of the complexity of the body/brain that supports the physical activity."

References

Appleby, M. (2014). Mind and matter: A critique of Cartesian thinking. In H. Atmanspacher & C. Fuchs (Eds.), *The Pauli-Jung conjecture and its impact today*. Exeter: Imprint Academic.

Atmanspacher, H. (2012). Dual-aspect monism à la Pauli and Jung. *Journal of Consciousness Studies, 19*(9–10), 96–120.

Bohm, D. (1952a). A suggested interpretation of the quantum theory in terms of hidden variables. *Physical Review, 85*, 166–179.

Bohm, D. (1952b). A suggested interpretation of the quantum theory in terms of hidden variables. *Physical Review, 85*, 180–193.

Bohm, D. (1990). A new theory of the relationship of mind and matter. *Philosophical Psychology, 3*, 271–286.

Bohm, D., & Hiley, B. (1993). *The undivided universe: An ontological interpretation of quantum mechanics*. London: Routledge and Kegan Paul.

Chalmers, D. (2010). *The character of consciousness*. Oxford: Oxford University Press.

Condliffe, J. (2017). *DeepMind's groundbreaking AlphaGo Zero AI is now a versatile gamer*. MIT Technology Review (6 Dec 2017).

Crick, F., & Koch, C. (1995a). Are we aware of neural activity in primary visual cortex. *Nature, 375*, 121–123.

Crick, F., & Koch, C. (1995b). Consciousness and neuroscience. *Cerebral Cortex, 8*, 97–107.

Crick, F., & Koch, C. (2003). A framework for consciousness. *Nature Neuroscience, 6*, 119–126.
de Barros, A., & Oas, G. (2017). Can we falsify the consciousness-causes-collapse hypothesis in Quantum Mechanics? *Foundations of Physics, 47*(10), 1294–1308.
Dehaene, S., Lau, H., & Kouider, S. (2017). What is consciousness and could machines have it? *Science, 358*, 486–492.
Goldstein, S. (2010). Bohmian mechanics and quantum information. *Foundations of Physics, 40*, 335–355.
Haven E., & Khrennikov, A. (2013). *Quantum social science*. Cambridge: Cambridge University Press.
Haven, E., & Khrennikov, A. Y. (2017). *The Palgrave handbook of quantum models in social science*. London: Springer – Palgrave MacMillan.
Hoffman, D. D. (2005). What do you believe is true even though you cannot prove it? Edge 2005 Annual Question. https://www.edge.org/response-detail/10930
Khrennikov, A. (2015). Quantum-like model of unconscious-conscious dynamics. *Frontiers in Psychology, 6*, 997. https://doi.org/10.3389/fpsyg.2015.00997
Khrennikov, A., & Haven, E. (2013). Physics goes social: How behaviour obeys quantum logic. *New Scientist, 219*, 26–27.
Nelson, E. (1966). Derivation of the Schrödinger equation from Newtonian mechanics. *Physical Review, 150*, 1079–1085.
Nelson, E. (2014). Stochastic mechanics of particles and fields. In H. Atmanspacher, E. Haven, K. Kitto, & D. Raine (Eds.), *Quantum Interaction: 7th International Conference* (University of Leicester) (Lecture notes in computer science, Vol. 8369, pp. 1–5).
Pinker, S. (1997). *Can computer be conscious*. US News and World Report, 123 (7) (18 Aug 1997).
Plotnitsky, A. (2017). The real and the mathematical in quantum modeling: From principles to models and from models to principles. *Frontiers in Physics, 5*(19). https://doi.org/10.3389/fphy.2017.00019
Plotnitsky, A., & Khrennikov, A. (2015). Reality without realism: On the ontological and epistemological architecture of quantum mechanics. *Foundations of Physics, 25*(10), 1269–1300.
Pylkkänen, P. (2015). Quantum theory, active information and the mind-matter problem. In E. Dzhafarov, S. Jordan, R. Zhang, & S. Cervantes (Eds.), *Contextuality from quantum physics to psychology*. (Advanced series on mathematical psychology, Vol. 6, pp. 325–334). River Edge: World Scientific.
Pylkkänen, P. (2017). Is there room in quantum ontology for a genuine causal role of consciousness. In E. Haven & A. Y. Khrennikov (Eds.), *The palgrave handbook of quantum models in social science* (pp. 293–317). London: Springer – Palgrave MacMillan.
Reginatto, M. (1998). Derivation of the equations of nonrelativistic quantum mechanics using the principle of minimum Fisher information. *Physical Review A, 58*(3), 1775–1778.
Skokowski, P. (2004). *Review on: G. Rosenberg: A place for consciousness: Probing the deep structure of the natural world*. Oxford University Press. Notre Dame Philosophical Reviews (7 Oct 2005).
Stapp, H. (1996). The hard problem: A quantum approach. *Journal of Consciousness Studies, 3*(3), 194–210.
Teasdale, G., & Jennett, B. (1974). Assessment of coma and impaired consciousness: A practical scale. *Lancet, 2*, 81–84.
Teasdale, G., & Jennett, B. (1976). Assessment and prognosis of coma after head injury. *Acta Neurochirurgica, 34*, 45–55.
Tononi, G. (2004). An information integration theory of consciousness. *BMC Neuroscience, 5*, 42.
Tononi, G., & Koch, C. (2008). The neural correlates of consciousness: an update. *Annals of the New York Academy of Science, 1124*, 239–261.
van Gulick, R. (2014) *Consciousness*. Stanford Encyclopedia of Philosophy. https://plato.stanford.edu/entries/consciousness/
Velmans, M. (1991). Is human information processing conscious? *Behavioral and Brain Sciences, 14*(4), 651–668.

Chapter 13
Panpsychism and Quantum Mechanics: Explanatory Challenges

Carlos Montemayor

Abstract This paper argues against a version of panpsychism that provides an interpretation of quantum mechanics, by appealing to phenomenal consciousness in order to address difficulties concerning measurements and observations. The challenge presented here is that phenomenal consciousness is neither necessary nor explanatory regarding issues in quantum mechanics. Rather, it is attention and access to information through epistemic constraints, such as rationality, that is necessary and explanatory. This suggests that if one takes mentality to be an essential ingredient in the interpretation of quantum mechanics, then mentality must be defined in terms of access, rather than phenomenal, consciousness.

13.1 Introduction

Panpsychism is an ancient and central view in the history of philosophy, with a long and venerable line of thinkers from various traditions. Recently, in the wake of several challenges showing the scientific intractability of consciousness, panpsychism and other non-physical theories of the mind have returned, reinvigorated after decades of dominance of materialist and physicalist (largely reductive) approaches. But panpsychism is a particularly flexible view, with many versions, some of which resemble materialist monist views, while others come closer to dualist metaphysical approaches. I shall not attempt to present (let alone criticize) panpsychism in all its complexity—a task which demands a book-length treatment. I shall rather focus on a specific claim of one of the influential versions of contemporary panpsychism, namely the claim that panpsychism might provide the best scientific explanation of consciousness, based on the best available theory of reality: quantum mechanics.

C. Montemayor (✉)
Department of Philosophy, San Francisco State University, San Francisco, CA, USA
e-mail: cmontema@sfsu.edu

J. Acacio de Barros, C. Montemayor (eds.), *Quanta and Mind*, Synthese Library 414, https://doi.org/10.1007/978-3-030-21908-6_13

A crucial feature of contemporary panpsychism is that it provides a unique perspective on the hard problem of consciousness, namely the problem of why should any cognitive function, semantic content or piece of information be accompanied with the distinct phenomenal character of subjective experience (Chalmers 1995). In fact, prominent proponents of panpsychism argue that panpsychism might provide the best explanation of consciousness, defined as the qualitative character of experience that resist all kinds of scientific reduction or explanation, while avoiding the problems surrounding emergence (Nagel 1979). The hard problem is itself an intricate issue. Here I shall assume it is a legitimate problem, one that demands solution, and focus on the merits of panpsychism as a scientific explanation of consciousness that would solve the hard problem. Contemporary panpsychists claim that their metaphysical explanation of consciousness provides a superior solution to the hard problem than physicalists or dualists views. Some versions of panpsychism use a mind-dependent interpretation of quantum mechanics to support this claim. My main focus will be this allegedly explanatory claim. I assess what this appeal to quantum mechanics really amounts to, and whether or not it efficiently bolsters the scientific credentials of panpsychism.

Since panpsychism is currently presented as a metaphysical view of consciousness that explains the nature of consciousness better than alternative views, it must articulate an explanation of its main claims in scientific terms. A prominent panpsychist argument appeals to the fundamentality of phenomenal consciousness at the quantum (micro-physical) level. This claim about fundamentality concerns the irreducibility of phenomenal consciousness (as a set of phenomenal properties or as a global and uniform, fundamental aspect of reality) to the structural and dynamical properties of matter and physical reality. The fundamentality of consciousness can be elucidated in terms of the following commitments:

Metaphysical-Primitivism Consciousness cannot be explained by extrinsic relations that inform us about structure and dynamics, so it must be an intrinsic aspect of reality.

Here it is useful to draw an analogy with matter: physics tells us what matter does, not what it is (see Russell 1927; Goff 2017). This is a perfectly valid and intelligible metaphysical claim, but is it an explanation of consciousness? By itself, this claim is compatible with many explanations that may not reduce consciousness to a specific understanding of physics, in terms of structure and dynamics, but which nonetheless are incompatible with the claim that consciousness is a primitive feature of the universe that cannot be explained in any physicalist terms (Stoljar 2010). Thus, it might be true that consciousness cannot be explained by structure and dynamics and yet, that it might be explained physically by some suitable definition of "physical." An alternative, which is the one examined here, is that redefining "consciousness" might help formulate this metaphysical claim. This has the advantage that a redefinition of consciousness might appeal to some notion of information. For primitivism to be more *explanatory*, rather than axiomatic, it must appeal, somehow, to a scientifically verifiable notion:

Information-Primitivism Consciousness must be a highly integrated informational system, not reducible to standard, structural and dynamical, accounts of information.

There is a theory that is explicitly addressing these two commitments through a scientific approach, namely the Integrated Information Theory (Tononi et al. 2016). There is no space to delve into the details of this ambitious and controversial theory here. What is central is that some notion of information must be in place because, otherwise, a difficulty emerges: how can something so informative (phenomenal consciousness) be so indescribable and mysterious in terms of explanation and information?[1] As mentioned, some scientific approaches to metaphysical primitivism appeal to quantum mechanics. To keep the discussion manageable, I focus on two quantum mechanical versions of contemporary panpsychism, and leave aside the thorny issue of information primitivism.[2] The main point of including information in the discussion is that an alternative definition of the type of mentality involved in quantum mechanics might help the panpsychist, but at the cost of abandoning their explanation of phenomenal consciousness.

13.2 Quantum Panpsychism: First Version

One version of panpsychism that appeals to quantum-fundamentality postulates that phenomenal consciousness is more fundamental than physical reality, which depends on the mental. There are various versions of this claim, but I shall consider a generic version of it.[3] An intuition that supports the claim that consciousness is more fundamental is that it seems, at least epistemically, more primitive, because we know it immediately, or a priori.[4] Relying on this intuition in the context of quantum mechanics, the claim is that all fundamental phenomena in quantum mechanics are dependent on the observations of a mind (more precisely, a phenomenally conscious mind). There is an intrinsic, "poised realm," in which a fundamentally mental and intrinsic reality (which we know directly) determines the mind-dependent probabilities that constitute the interactions in quantum mechanics, upon observation.

[1] For several important solutions to problems confronting panpsychism as an explanation of consciousness, including mapping and combination problems, a very useful volume is Brüntrup and Jaskolla (2017).

[2] See de Barros et al. (2017) for problems concerning the axioms of IIT with respect to contextuality.

[3] The closest analog of this formulation in interpretations of quantum mechanics is perhaps Stapp (2007).

[4] This claim is related, although in ways that require argumentation, to the thesis that consciousness is relevant to understanding the intrinsic nature of reality because we know it better than matter. See, for instance, Whitehead (1933); Strawson (2003); Goff (2017).

Even if this controversial interpretation of quantum mechanics were accepted, it can be interpreted without phenomenal consciousness. This has advantages that have not been discussed at length in the literature on panpsychism. I present two alternative interpretations supported by empirical and theoretical considerations. First, since observation and prediction are best understood in terms of *attention* (and rational inference), there is no reason to accept the view that phenomenal consciousness is necessary at the fundamental level. One then needs to independently worry about the mystery of phenomenal consciousness. But nothing in the mind-dependent version above demands phenomenal consciousness. All that is required is a rational mind with powers of observation, and a mind with attention capacities suffices. In other words, this mentalistic view is compatible with not-phenomenally conscious observational detection. A further advantage of interpreting mentality in terms of attention is the intrinsically normative nature of observations and rationality in terms of epistemic constraints, related to access consciousness (Montemayor and Haladjian 2015) and attention (Fairweather and Montemayor 2017). In fact, since mentality is fundamental reality but presumably it cannot be exclusively subjective (the unique domain of a single perspective), then access consciousness seems more adequate, as accessible information; access to information serves a more explanatory role than aspects of what it is like to be me (or you). The mental *poised realm*, presumably, cannot be what it is like to be neither of us. So if it is construed as conscious information it should be "access" conscious information (Block 1995), and this can be done through attention without phenomenal consciousness (Montemayor and Haladjian 2015).

Second, and for similar reasons, since there are grounds to doubt that only human consciousness is fundamental, a non-anthropocentric approach demands that rationality, rather than phenomenal consciousness, should be fundamental. This also suggests that attention is better suited to play this rational role. Consciousness is not necessary to defend the first, mind-fundamental, interpretation of quantum mechanics. In fact, as I am about to explain, the scientifically confirmed dissociation between consciousness and attention speaks against this interpretation. So even if one favors a kind of fundamental panpsychism in quantum mechanics, this view is best understood in terms of pan-attentional or pan-intentional mentality, rather than phenomenally conscious, subjective experience. In particular, the kind of rational "noticing" in terms of yes or no answers, or measurements, is best understood in terms of not necessarily conscious attention (which could be pre-conscious or unconscious—pre-phenomenal or not related to the phenomenal at all). Thus, in Stapp's interpretation (based on von Neumann's), the cosmic mind is best understood in terms of attention. In fact, attention and intentional action are emphasized by Stapp (1999). (The fundamentality of the minds of observers and nature's "choices" is discussed in [2015]; see also Montemayor 2016).

Attention routines for the detection of a feature or object, with a yes or no answer output, are very ancient and well documented across species (Haladjian and Montemayor 2015). The dissociation between consciousness and attention is also very well confirmed. For instance, an argument from evolution supports it (Montemayor and Haladjian 2017). Perceptual systems evolved from basic to complex forms of processing, and they can be defined in terms of intentionality (the way in which mental representations are *about* objects and features in the world). If perceptual systems evolved, then intentionality also evolved. Therefore, some forms of intentionality involved in types of attention routines are more ancient and need to be understood in terms of attention. But did phenomenal consciousness evolve as early as attention? This is not clearly the case (see Montemayor and Haladjian 2017 for details). Crucially, this dissociation is neutral with respect to current theories of consciousness, because all of them entail some degree of dissociation between consciousness and attention (Montemayor and Haladjian 2015).[5]

13.3 Quantum Panpsychism: Second Version

An alternative version of panpsychism postulates the fundamentality of proto-phenomenal properties, which are as fundamental as the properties of mass or charge, and somehow constitute phenomenal consciousness without depending on physical properties (consciousness is an intrinsic aspect of reality, but it is "on a par" with physical reality). This version shares the commitment of the previous view to the primitive and intrinsic aspects of consciousness, but denies its fundamentality, allowing for a *dual aspect* interpretation, compatible with Russellian monism. Although this process of integration from the primordial realm into conscious beings generates notorious composition problems, this view has the advantage that it avoids the controversial assumption concerning the priority of the mental.

However, and similarly to the previous case, phenomenal consciousness is not necessarily playing an *explanatory* role in this interpretation of quantum mechanics. Instead, other cognitive processes, such as attention and access to information, seem to be necessary in the formulation of the transition from proto-mental to mental, in quantum terms. Moreover, the compositional constraints seem to be operating at a different scale than the micro-level, in which case emergentism must be fundamentally assumed, with all the controversial aspects of emergentists approaches to causation.[6] Yet another possibility is that the same quantum principles operate in the mental and physical aspects of the monistic reality, but this dual aspect interpretation also demands access to information, rather than phenomenal subjectivity.

[5]Moreover, there are normative reasons for this dissociation. Attention, as mentioned, is best explained in terms of epistemic normativity and rationality, of the kind we expect in scientific measurements, but phenomenal consciousness is related to moral status and moral value (Montemayor and Haladjian 2015). Even in terms of brain anatomy, rational choices are linguistically framed and may require the frontal cortex and the attention areas, but it is quite controversial that phenomenal consciousness necessitates these areas.

[6]See Seager (2017) and Mørch (2014) for solutions to emergentist problems concerning this type of view. See Papineau (2001) for criticisms of any type of emergentism. See also Montemayor (2017) for reasons in favor of a purely physicalist interpretation of monism, based on considerations about information.

In any case, there seem to be two challenges to this approach concerning the explanatory role of phenomenal consciousness. First, in order to satisfy any type of compositional requirements, the mapping should involve epistemically constrained contents and the structural aspects of fundamental reality on which existence they depend. These epistemic constraints, again, are sufficiently satisfied by attention without phenomenal consciousness. In fact, there seems to be a lack of isomorphic mapping between phenomenal properties and anything fundamental at the quantum scale, or even at the "emergent" levels (or even some mapping relation—although isomorphism seems to be necessary).[7]

A second problem, also related to a different type of unification and composition difficulty, is the lack of straight-forward mapping between the indexical information of *subjectivity* required for phenomenal consciousness and anything fundamental at the quantum scale (or even at emergent scales). This is related to the issue mentioned before, which is that access consciousness and, in particular, attention, seems better suited to establish this mapping independently of the private and subjective character of the first person perspective, although admittedly it is controversial to fully separate the first person point of view from attention. Perhaps any type of panpsychism is incompatible with the kind of indexical information required for the perspective-dependent information characteristic of consciousness—a relation between *actual* indexical information and *possible* physical information. However, if some type of agency is required for a panpsychist quantum mechanics, attention suffices to provide such an account, in terms of epistemic agency without phenomenal consciousness (see Fairweather and Montemayor 2017 for such an account).

These mapping problems are also central difficulties for any phenomenal consciousness-based interpretation of the monistic universe that manifests in quantum mechanics. For this interpretation to work, macro-properties must be structurally dependent (structurally identical and presumably isomorphic) to the structure of micro-properties. Alternatively, they must strongly emerge on the micro-properties and *explain*, how quantum mechanics acquires the properties it has. But it is unclear what this means in terms of the specificity of semantic content, for instance, of the semantics of terms like 'pain' or 'red,' and the unique perspective of subjectivity, and how this subjective information maps to the mathematical structure of quantum mechanics.

Similar issues apply to questions concerning the right scale or level at which one can identify phenomenal properties objectively (a somewhat paradoxical kind of identification that is required for phenomenal consciousness to explain quantum mechanics, or at least to explain its relation to quantum mechanics). The brain, for instance, might be at the relevant scale for semantic and phenomenal mappings, but that scale might be too coarse for any quantum-level determinate information. Thus, this is not only an issue of identifying the right mappings at the right structural

[7]This problem also affects panprotopsychism as well, which is a view that weakens the metaphysical commitments of traditional panpsychism, because it doesn't take consciousness per se to be fundamental, but some other precursor to consciousness instead.

level, but also of providing an explanation of how could one in principle identify the right scale for interactions between the conscious and physical aspects of reality. It seems more manageable to assume that the mappings must be between a rationally constrained notion of mentality and physical aspects of reality, without assuming the involvement of phenomenal consciousness.

An additional difficulty is, how can emergentist accounts of panpsychism avoid problems related to supervenience and mental causation? For these accounts to qualify as genuinely panpsychist they must avoid commitments to composition principles from parts to wholes that appeal to an irreducible unification principle at the conscious level. The next sections present distinctions that might help clarify these different versions of panpsychism. These composition and mapping problems are central difficulties for panpsychism, widely discussed in the literature. What is relevant for our purposes is that panpsychism seems to lack an explanation of quantum mechanics, if it is based on phenomenal consciousness. Even if one endorses the claim that some type of mentality is fundamental to explain quantum mechanics, attention and access consciousness are much better suited to the explanatory task than phenomenal consciousness. As I proceed to explain, there are powerful reasons for this, based on differences in information.

13.4 An Informational Difference

At the core of the challenges described above is a fundamental difference between attention and consciousness regarding information. Attention is informative in a *computational* way; it halts after a routine and it integrates information in order to conclude the routine. It provides an answer to a question that can be publically assessed and epistemically evaluated. It also provides an interface for epistemic exchanges, rationality and inference, which can be understood in terms of a language with semantic values. Joint attention, for instance, is fundamental for collective epistemic achievements and the formation of collective epistemic motivations (Fairweather and Montemayor, Ch. 7).

Consciousness, by contrast, is informational in a *homeostatic* way; it never "halts," and its purpose is to provide engaging and empathic access (Haladjian and Montemayor 2016) to regulatory systems that are integrated vividly into the first person perspective. Biologically speaking, it has a fundamental kind of "grip," associated to subjectivity and the privacy of the first person point of view—phenomenal consciousness is *visceral*. It doesn't mean something to the individual in a way that is merely shared with others from an epistemic and rational perspective. Rather, it is shared only in so far as others have similar empathic and visceral reactions to their biology, from their personal point of view. Even if one grants that physicalism is not in a better position to explain consciousness than panpsychism, it does not follow that panpsychism is an *explanation* of consciousness or of anything science can study. Partly because of this difference in information, which has yet to be accounted for in detail in contemporary debates, panpsychism at this stage might be a collection of arguments and a priori claims, rather than a scientific explanation, or even the very beginnings of it. If, however, one interprets the irreducibly mental reality of panpsychism in terms of attention, an information theoretic approach might provide the beginnings of such an explanation.

But what about the "hard problem" of consciousness? Doesn't this problem show that physicalism is doomed to keep the main mystery of consciousness in the shelf of impossibly difficult and unsolvable problems? The main attraction of panpsychism is that it might solve this problem, as mentioned in the introduction. Identifying attention is easy, to a large extent because computational approaches seem to suffice as an explanation of the functions of attention (although purely computational approaches may be far from sufficient because of issues concerning motivation; see Fairweather and Montemayor 2017). That means that solving all the problems concerning the nature of attention, daunting as they might be, fall under the umbrella of "easy problems", compared to the enigma of phenomenal consciousness. Explaining consciousness is hard, and appealing to physicalist and functional approaches leaves the issue entirely untouched. Hence the scientific or explanatory motivation for panpsychism.

I provide a more positive take on the hard problem in the next section. Here I shall highlight that the hard problem does not show that any other non-physical theory is explanatory. That needs to be shown independently, through a theory of information or something analogous to a theory of the conscious mind. In addition, with respect to interpretations of quantum mechanics that appeal to consciousness, it is important to highlight the *sufficiency* of attention, as a type of "mind" that can explain quantum mechanics without necessarily appealing to phenomenal consciousness: consciousness is not necessary to explain what those interpretations want to explain. The mentality involved in these interpretations can be perfectly understood in terms of attention. Attention suffices for rationality, which is the main feature of the mental realm assumed in mind-dependent interpretations of quantum mechanics. Attention is actually the best way to understand the binary decision or choice related to measurements, associated with these interpretations.

Accordingly, the debate surrounding these interpretations of quantum mechanics should include other types of mentality for contrasting conscious with non-conscious attention. The presence of certain skills, used for identifying consciousness, might not be associated necessarily with consciousness, but with forms of attention that are not essentially conscious. Skills may be implemented, for instance, in artificially intelligent systems. This presents a major challenge for assessing consciousness based solely on these skills. Rationality or intelligence are not necessarily associated with phenomenal consciousness. The mental activities involved, for instance, in a decision to measure and the detection of an outcome (yes or no), is a strictly attentional process.

Nonetheless, the hard problem provides a very strong reason to favor panpsychism, particularly given the limitations of physicalism and the metaphysical intricacies of Cartesian-style dualism. It would be desirable to say more about how this problem could be addressed by using the distinction between consciousness and attention. The next section provides a sketch of how it could be possible to address this seemingly intractable problem.

13.5 The Meta-Problem and the Hard Problem

In a recent paper, David Chalmers considers a problem that is related to the hard problem in an *explanatory* way, and evaluates the possibility that solving this interrelated, meta-problem, might shed some light on the hard problem. These are the three problems one confronts in consciousness studies, according to Chalmers:

Easy Problems These are problems concerning the scientific explanation of behavioral functions, cognitive functions, intentionality, and intelligence (even Artificial Intelligence), dependent on the structure and dynamics of bio-chemical and non-biological informational systems. As mentioned previously, most of the problems concerning the explanation of attention would presumably fall under the umbrella of easy problems.

Hard Problem This is the metaphysical problem of why any type of cognitive function must be accompanied by subjective experience, with its unique phenomenal character? (Chalmers 1995, 1996). The hard problem is a key aspect of the arguments in favor of anti-physicalists views of consciousness, including panpsychism.

Meta-Problem This is the epistemic problem of *why do we think* consciousness is hard to explain (Chalmers 2018)? Unlike the previous two problems, this problem requires a psychological and epistemic explanation which could be provided in terms of experimental psychology and findings in cognitive science.

I propose three tentative answers to the Meta-Problem. We think explaining consciousness is hard because: (a) phenomenal consciousness is not computationally informative (or more precisely, not merely computational). It is more integrative, in a visceral and biologically-based type of way—by means of homeostasis and empathic engagement; (b) the lack of "halting functions" for consciousness, namely functions with specific solutions to problems that are delivered in terms of information detection and outputs, makes us think that consciousness is irreducible to cognitive function and physical structures and patterns, including behaviors; and (c) the homeostatic relation to empathic, integrative states, makes us think that consciousness is much more *unique* than computational information. These three related proposals show that we believe consciousness is hard to explain because it is non-computational and uniquely valuable, and this is compatible with the experimentally confirmed dissociation between consciousness and attention: attention is epistemically valuable and phenomenal consciousness is morally valuable because of its uniqueness and irreproducibility (see Montemayor and Haladjian 2015, 2017).

Perhaps the problem that needs to be tackled now is the meta-problem. Once it is solved, along the lines suggested here, we can shed light on the uniqueness and value of phenomenal consciousness, which is a topic that needs to be explored in much more detail. The meta-problem can be solved in an informational way and science can certainly help solving it, partly by analyzing in depth the distinction between consciousness and attention. The explanation will include considerations about evolution, brain anatomy, structures for empathy and social behavior (Haladjian and Montemayor 2015). Meanwhile, interpretations of quantum mechanics that appeal to the mind can rely on the rational and epistemically constrained functions of attention (a type of attentional panpsychism, rather than a panpsychism of the phenomenally conscious mind).

These proposed solutions to the meta-problem have the advantage of addressing the notion of *creature consciousness* in a direct and explanatory way. It has been proposed that the function of the brainstem is to provide a biologically centered basis for what is it like to experience contents from the perspective of an organism—this is the idea behind having consciousness without a cerebral cortex (Merker 2007). This function is best understood as *creature consciousness*, rather than access or phenomenal consciousness (Block 1995, 1997), and it can be identified as the condition that is necessary for the contents of phenomenal consciousness to be vivid and visceral. In combination with the distinction between consciousness and attention, one can argue that creature consciousness is a necessary condition for phenomenal (state) consciousness, but not for global access to information, including various forms of epistemically guided attention (e.g., attention that provides perceptual and inferential knowledge and justification). If creature consciousness is necessary for phenomenal consciousness, however, then panpsychism is false.

Finally, since what is at stake in interpreting the kind of mentality involved in interpretations of quantum mechanics concerns types of information, computational-rational and homeostatic-conscious, more needs to be done in order to understand consciousness in terms of information. Until we understand the nature of information better, it is premature to choose any type of panpsychism over physicalism. In particular, we need to understand the role of contextuality in essentially indexical information, which eliminates possibilities in a unique and causally anchored way. In the case of homeostatic information, actual (versus possible) information is anchored in a biologically constrained way. The issue of indexical information, associated with the actual (here, now) and the personal (I) versus the possible (the many-worlds interpretation) is already a pressing issue in physics and quantum mechanics. This topic is even more pressing for interpretations of quantum mechanics that appeal to the mind.

13.6 Conclusion

Panpsychism comes in different versions, and it covers an intricate and broad set of views. As mentioned in the introduction, it is a powerful approach to many difficult topics in philosophy of mind and metaphysics. The focus of this paper has been to argue against a very specific claim of a kind of panpsychism that appeals to quantum mechanics, namely the claim that consciousness helps explain key problems in quantum mechanics, including the measurement problem. The challenge presented here is that consciousness, even if one accepts these controversial interpretations, is neither necessary nor explanatory of issues in quantum mechanics. Rather, it is attention and access to information through epistemic constraints, such as rationality, that is necessary and explanatory.

Suppose that panpsychism is framed according to the empirically and theoretically supported distinction between consciousness and attention, by describing the omnipresent mind in terms of attention and rationality rather than phenomenal consciousness and subjective experience. Wouldn't that leave the hard problem unaddressed and unsolved? I argued that perhaps a good way of approaching the hard problem is by first solving what Chalmers recently called the "meta-problem." The main strategy I proposed to solve this latter problem is precisely in terms of the distinction between consciousness and attention. This might provide both a more plausible panpsychist interpretation of quantum mechanics and an alternative approach to the hard problem.

Ultimately, if phenomenal consciousness is to be explained scientifically, it must presumably be explained in terms of information theory. If there is phenomenal information that cannot be reduced to physical or computational information then, presumably, a theory of information can explain how this is the case. Regardless of how this issue is solved, it is important to consider that theorists in all disciplines define "the mind" differently. Moving forward, it is important to take into consideration the dissociation between phenomenal consciousness and attention in these debates. Among other reasons, as explained before, this distinction may help elucidate the epistemic value of mental processes associated with attention and rationality and the moral value of mental processes associated with phenomenal consciousness. The latter is not essential for interpretations of quantum mechanics.

References

Block, N. (1995). On a confusion about a function of consciousness. *Behavioral and Brain Sciences, 18*(2), 227–247.

Block, N. (1997). On a confusion about a function of consciousness. In N. Block, O. J. Flanagan, & G. Güzeldere (Eds.), *The nature of consciousness: Philosophical debates* (pp. 375–415). Cambridge, MA: MIT Press.

Brüntrup, G., & Jaskolla, L. (Eds.). (2017). *Panpsychism: Contemporary perspectives*. New York: Oxford University Press.

Chalmers, D. J. (1995). Facing up to the problem of consciousness. *Journal of Consciousness Studies, 2*(3), 200–219.

Chalmers, D. J. (1996). *The conscious mind: In search of a fundamental theory*. New York: Oxford University Press.

Chalmers, D. J. (2018). The meta-problem of consciousness. *Journal of consciousness studies, 25*(9–10), 6–61.

de Barros, J. A., Montemayor, C., & de Assis, L. P. G. (2017). Contextuality in the integrated information theory. In J. A. de Barros, E. Pothos, & B. Coecke (Eds.), *Quantum interaction: Lecture notes in Computer Science (10106)* (pp. 57–70). Cham: Springer.

Fairweather, A., & Montemayor, C. (2017). *Knowledge, dexterity, and attention*. New York: Cambridge University Press.

Goff, P. (2017). *Consciousness and fundamental reality*. New York: Oxford University Press.

Haladjian, H. H., & Montemayor, C. (2015). On the evolution of conscious attention. *Psychonomic Bulletin and Review, 22*(3), 595–613.

Haladjian, H. H., & Montemayor, C. (2016). Artificial consciousness and the consciousness-attention dissociation. *Consciousness and Cognition, 45*, 210–225.

Merker, B. (2007). Consciousness without a cerebral cortex: A challenge for neuroscience and medicine. *Behavioral and Brain Sciences, 30*(1), 63–81.

Montemayor, C. (2016). Commentary on Stapp. In S. O'Nualláin (Ed.), *Dualism, platonism and voluntarism: Explorations at the quantum, mesoscopic and symbolic neural levels* (pp. 193–204). Cambridge: Cambridge Scholars Publishing.

Montemayor, C. (2017). The problem of the base and the nature of information. *Journal of Consciousness Studies, 24*(9–10), 91–102.

Montemayor, C., & Haladjian, H. H. (2015). *Consciousness, attention, and conscious attention*. Cambridge, MA: MIT Press.

Montemayor, C., & Haladjian, H. H. (2017). Perception and cognition are largely independent, but still affect each other in systematic ways: Arguments from evolution and the consciousness-attention dissociation. *Frontiers in Psychology, 8*, 40. Edited by Athanassios Raftopoulos and Gary Lupyan, 8.

Mørch, H. H. (2014). *Panpsychism and causation: A new argument and a solution to the combination problem*. Ph.D. thesis, University of Oslo.

Nagel, T. (1979). Panpsychism. In *Nagel's mortal questions* (pp. 181–195). Cambridge: Cambridge University Press.

Papineau, D. (2001). The rise of physicalism. In C. Gillett & B. M. Loewer (Eds.), *Physicalism and its discontents* (pp. 2–36). Cambridge: Cambridge University Press.

Russell, B. (1927). *The analysis of matter*. London: George Allen and Unwin.

Seager, W. E. (2017). *Panpsychism infusion* (Brüntrup, & Jaskolla, Eds.), pp. 229–248.

Stapp, H. (1999). Attention, intention, and will in quantum physics. *Journal of Consciousness Studies, 6*(8/9), 143–164.

Stapp, H. (2007). *Mindful universe*. Berlin: Springer.

Stoljar, D. (2010). *Physicalism*. New York: Routledge.

Strawson, G. (2003). Real materialism. In L. Antony & N. Hornstein (Eds.), *Chomsky and his critics*. Oxford: Blackwell.

Tononi, G., Boly, M., Massimini, M., & Koch, C. (2016). Integrated information theory: From consciousness to its physical substrate. *Nature Reviews: Neuroscience, 17*, 450–461.

Whitehead, A. N. (1933/1961). *Adventures of ideas*. New York: Macmillan.

Chapter 14
Quantum Theory and the Place of Mind in the Causal Order of Things

Paavo Pylkkänen

Abstract The received view in physicalist philosophy of mind assumes that causation can only take place at the physical domain and that the physical domain is causally closed. It is often thought that this leaves no room for mental states *qua* mental to have a causal influence upon the physical domain, leading to epiphenomenalism and the problem of mental causation. However, in recent philosophy of causation there has been growing interest in a line of thought that can be called causal anti-fundamentalism: causal notions cannot play a role in physics, because the fundamental laws of physics are radically different from causal laws. Causal anti-fundamentalism seems to challenge the received view in physicalist philosophy of mind and thus raises the possibility of there being genuine mental causation after all. This paper argues that while causal anti-fundamentalism provides a possible route to mental causation, we have reasons to think that it is incorrect. Does this mean that we have to accept the received view and give up the hope of genuine mental causation? I will suggest that the ontological interpretation of quantum theory provides us both with a view about the nature of causality in fundamental physics, as well as a view how genuine mental causation can be compatible with our fundamental (quantum) physical ontology.

14.1 Introduction

The received view in physicalist philosophy of mind states that causation can only take place at the physical level because everything that happens in the world is ultimately determined by the laws of physics and the physical domain is causally closed (cf. Sundström and Vassen 2017). This implies that non-physical entities cannot have any physical effects. If mind (whether conscious or unconscious) is

P. Pylkkänen (✉)
Department of Philosophy, History and Art Studies, University of Helsinki, Helsinki, Finland

Department of Cognitive Neuroscience and Philosophy, University of Skövde, Skövde, Sweden
e-mail: paavo.pylkkanen@helsinki.fi

J. Acacio de Barros, C. Montemayor (eds.), *Quanta and Mind*, Synthese Library 414, https://doi.org/10.1007/978-3-030-21908-6_14

taken to be non-physical (as is often done), then it cannot have any physical effects, and we end up with epiphenomenalism, the view that mental properties exist but have no causal influence upon the physical world. Epiphenomenalism is widely thought to be an unsatisfactory view and there have been many attempts to avoid it (see Robb and Heil 2018). In this paper I will discuss two ways to challenge the received view, and to save mental causation.

The first is provided by causal anti-fundamentalism, a view which states that causation is not part of the fundamental physical ontology of the world. If causal anti-fundamentalism is correct, then the received view is mistaken. Moreover, it has been suggested that causal anti-fundamentalism is compatible with the idea that fundamental physical facts ground higher-level causal facts, including those involving consciousness (Blanchard 2016: 256). Conscious experiences could then be seen as local events which determine events in their future.

A second way to challenge the received view is opened up by the ontological interpretation of quantum theory (Bohm and Hiley 1987, 1993). This interpretation suggests that the wave function in quantum theory describes a new type field which contains active information, which latter is a fundamental, causal factor organizing the motion of physical particles. Bohm (1990) proposed that by extending this quantum ontology in a natural way one can show how mental properties can influence matter. If Bohm's proposal is correct the received view is either mistaken, or else the concept "physical" in the received view needs to be extended to include active information and possibly mind/consciousness (which are traditionally often taken to be non-physical entities).

14.2 Causal Anti-fundamentalism

Philosophers of mind and neuroscientists typically assume that we have a clear understanding of the nature of causality in the physical domain. But, in fact, there is a venerable tradition, dating back to Russell's (1913) causal anti-fundamentalism, arguing that causal notions can play no legitimate role in how physics represents the world (Frisch 2012; Price and Corry 2007). Causal anti-fundamentalism has been succinctly summarized in a recent workshop description by Sundström and Vassen (2017):

> ...causal notions cannot play a role in physics, because the fundamental laws of physics are radically different from causal laws. Causal laws typically describe how local events determine events in their future; for example, a causal law can connect smoking to later occurrences of cancer. By contrast, physical laws connect the entirety of physical reality in a time-symmetric manner: the entire state of the universe at a certain time equally determines the relative past and the future of the universe. It therefore appears reasonable to situate causation in the higher levels of science where local events are studied in a time-directed manner, such as biology and economy.

So, causal anti-fundamentalism is the thesis that causation is not part of the fundamental physical ontology of the world. If true, this entails that causation is not as central a feature of the world as ordinarily thought. But as was noted above, it has been suggested that causal anti-fundamentalism is compatible with the existence of causal facts as non-fundamental features of the world grounded in fundamental physical facts (Blanchard 2016: 256 cf. Cartwright 1979). I suggest that causal facts involving conscious experiences can also be seen as such features. This suggests a view in which conscious experiences can be seen as local events which determine events in their future (in a similar manner as with causal laws in special sciences such as biology and economy). Physics (and the causal closure of the physical domain) would no longer be an obstacle to mental causation. On the contrary: fundamental physical facts would now *ground* the causal facts involving conscious experiences![1]

Let us summarize our discussion so far. According to the received physicalist view there is no room for mental causation, because all goings on in the world are ultimately determined by physics and physics itself is causally closed. In contrast, according to the view inspired by causal anti-fundamentalism the physical world is not governed by any fundamental relation of cause and effect. But one can still hold that fundamental physical facts ground higher-level causal facts, including those involving consciousness. Conscious experiences can then be seen as local events which determine events in their future. Mental causation is possible!

Is this too good to be true? It is true that causal anti-fundamentalism, when suitably interpreted, gives us mental causation in some sense. But it also implies that causal facts (including those involving conscious experiences) are non-fundamental. One might worry whether this is too deflationist a view of (mental) causation. Also, causal anti-fundamentalism is admittedly weird, given the way we have traditionally been thinking about the laws of physics. Is it really the case that the physical world is not governed by any fundamental relation of cause and effect?

14.3 Is Causal Anti-fundamentalism Correct?

Remember that causal anti-fundamentalism claims that the fundamental laws of physics are radically different from causal laws. Allegedly, one important difference is that while causal laws typically describe how *local events* determine events in their future, physical laws connect the *entirety of physical reality*. But do physical laws really connect the entirety of physical reality? In the spirit of scientific metaphysics, let us here consider how a physicist might answer the question. When discussing causality and chance in modern physics David Bohm proposed in 1957

[1]Metaphysical grounding is, of course, a subtle topic in contemporary metaphysics, and we will not here enter into a discussion about what exactly it might mean when one says that fundamental physical facts ground causal facts. But see Bliss and Trogdon 2016. An additional challenge is to spell out how physical facts could ground non-physical, conscious facts (thanks to Tuomas Tahko for pointing out this challenge).

that there is no known fundamental physical law that would be able to take into account the complete state of the world:

> ... every real causal relationship, which necessarily operates in a finite context, has been found to be subject to contingencies arising outside the context in question. (1957/1984: 3)

If Bohm is correct, perhaps the fundamental laws of physics are *not* radically different from causal laws, and there is causation in fundamental physics after all (for a brief discussion of Bohm's view, see Andersen et al. 2018).

What about time-symmetry? Another reason why causal anti-fundamentalism claims that the fundamental laws of physics are radically different from causal laws has to do with time-symmetry. Causal laws typically describe how local events *determine events in their future*. But physical laws connect the entirety of physical reality in a *time-symmetric manner*. The idea is that the entire state of the universe at a certain time equally determines the relative past and the future of the universe.

Let as again consider the issue in the light of Bohm's thinking about physics. When Bohm in his 1951 text-book *Quantum theory* discusses time-symmetry he notes that classical theory is prescriptive and not causal. He points out, consistently with causal anti-fundamentalism, that in classical physics the idea of forces as causes of events became unnecessary and almost meaningless.

> [this is so] because *both the past and the future* of the entire system are determined completely by the equations of motion of all the particles, coupled with their positions and velocities at any one instant of time. Thus we can no more say that the future is caused by the past than we can say that the past is caused by the future. Instead, we say that the motion of all particles in space and time is *prescribed* by a set of rules, i.e. the differential equations of motion, which involve only these space-time motions alone. (1951: 151).

But what about the quantum theory? Here Bohm makes a point which seems to challenge causal anti-fundamentalism in a profound way:

> Whereas classical theory can be expressed in terms of a set of prescriptive rules relating space-time motions at different times, *quantum theory cannot be so expressed.* Energy and momentum (and therefore, the causal factors) cannot be eliminated in terms of velocities and positions of the component particles. The quantum theoretical concept of causality, therefore, differs from its classical counterpart. It must necessarily describe the relationships between space-time events as being "caused" by factors existing within matter (i.e., momenta). These are on the same fundamental and not further analyzable footing as that of space and time themselves. (1951: 157)

Bohm acknowledges that in quantum theory these causal factors control only a statistical trend in the course of space-time events. But he notes further that *it is just this property of incomplete determinism that prevents the causal factors from becoming redundant.* This, he says, gives a real content to the concept of causality in quantum theory.

So, if Bohm is correct, causal anti-fundamentalism is incorrect and there is causation in the physical world. This so, first of all, because each known fundamental physical law operates in a finite context, and secondly because in quantum theory, energy and momentum (and therefore, the causal factors) cannot be eliminated in terms of velocities and positions of the component particles (unlike in Newtonian physics).

Note, however, that what we have just reported above comes from Bohm's 1951 textbook *Quantum theory* where he was trying to explicate a version of the usual, "Copenhagen" interpretation (his explication is fairly similar to the approach of Wolfgang Pauli). But as is well known, Bohm himself published an alternative interpretation of quantum theory in 1952. This interpretation (or "the Bohm theory") assumed that an electron has simultaneously a well-defined position and velocity and is guided by a new type of field (mathematically described by the wave function Ψ). According to the usual interpretation of quantum theory, the wave function Ψ does not describe an individual quantum system directly. Rather, Ψ describes our *knowledge* of the quantum system to be observed (typically in terms of probabilities). In contrast, according to Bohm's 1952 theory, Ψ describes an objectively real field, guiding a particle such as an electron.

What is the role of causality in the Bohm theory? A minimalist version of the theory (called "Bohmian mechanics") expresses quantum theory as a "first-order" theory in terms of velocities (see Goldstein 2013). As far as I can judge, this is similar to classical physics in that causality (energy, momentum) can be eliminated if one wishes. In some versions of Bohmian mechanics the wave function is assumed to be law-like. So it seems that causality can be eliminated just as in classical physics![2] However, Bohm and Hiley developed Bohm's 1952 theory into another direction ("the ontological interpretation"), where the role of the so called quantum potential energy is important. In this picture it may be possible to retain energy as irreducible and thus "genuine" causality in quantum mechanics.[3]

The above suggests that there are certain ironies in the question quantum theory and causality. It seems that the usual interpretation of quantum theory implies that (statistical) causality is irreducible, ironically, *because of* the uncertainty principle. However, recent developments in the interpretation of quantum theory (e.g. Bohmian mechanics) may eliminate causality from fundamental physics (ironically, *because of* determinism, just as happened in Newtonian physics). But there is the possibility that in Bohm and Hiley's ontological interpretation, with its emphasis on quantum potential energy, we can retain genuine causality.

Let us now return to the issue of mental causation. We have noted that if Bohm and Hiley's interpretation is correct, and includes an irreducible form of quantum potential energy, then there is causation in the fundamental physical level and causal anti-fundamentalism is incorrect. But how can we then have mental causation? Doesn't the principle of the causal closure of the physical domain apply to the Bohm-Hiley scheme as well, thus leaving no room for mental properties *qua* mental to have an effect upon the physical?

---

[2]As I am not a physicist, I offer these proposals tentatively, to be discussed in more detail by those physicists specialized in Bohmian mechanics.

[3]Again, I am offering this proposal tentatively as a philosopher, to be discussed in more detail by physicists.

Here I suggest that we consider Bohm and Hiley's (1993) suggestion that the ontological interpretation can be extended. So let us move on to examine the ontological interpretation and the way its extension might allow for mental causation.

14.4 Extending the Ontological Interpretation of Quantum Theory

According to the ontological interpretation of quantum theory quantum processes are guided by a field containing active information (described by the wave function Ψ, expressed in terms of the quantum potential). This involves a new type of causation which we may call informational causation. This information is radically holistic – it enables non-locality and irreducible objective wholeness of a many-body system.

It is important to realize that Bohmian active information is not Shannon information. The idea is that the form of the quantum field (described by the wave function) enfolds information about the environment (e.g. slits), and this information then literally IN-FORMS the movement of the particle through the quantum potential. The information is potentially active everywhere where the quantum potential non-zero, but actually active only where the particle is. Without such actual activity of "in-forming" the information would have no causal powers. So in this picture there needs to be actually active information for there to be any causally efficacious information at all.

Bohm himself (1990) proposed that active information can be seen as a "primitive mind-like quality" of elementary particles, suggesting a view that we may call Bohmian panprotopsychism (see Pylkkänen forthcoming). The key principle here is that mental processes involve "activity of form" rather than "activity of substance". When you read the newspaper, you do not need to eat the paper, you abstract the form that is carried or enfolded in the movement of light waves. That form, when taken up by the nervous system and interpreted can give rise to a conscious experience of meaning of the information. Analogously, the electron is not pushed and pulled by the quantum wave. Rather, it is able to respond to the form of the quantum wave. It is "mind-like" in this sense. Note that Bohm thought it obvious that an electron is not (phenomenally) conscious (Bohm 1990). But we could ask whether the electron is in some sense "perceiving" (unconsciously) its environment via the quantum field. While many are still likely to see pan(proto)psychism as "a complete myth, a comforting piece of utter balderdash" (McGinn 2006: 93), it has become a subject of intense discussion in contemporary philosophy (see Strawson 2006; Goff et al. 2017; Pylkkänen forthcoming).

How does mental causation work in Bohmian quantum ontology? Bohm suggested that it is natural to extend the quantum ontology. Just as there is a quantum field that informs the particle, there can be a super-quantum field that informs the first order quantum field, and so on (Bohm and Hiley 1993: 379–381). Let us further assume that the information contained in our mental states is a certain part of this hierarchy of fields. Through the hierarchy, mental states could then guide material processes, by reaching the quantum field of the particles and fields in the brain (Bohm 1990).

How exactly might such" quantum mental causation" work? There are currently a number of different proposals regarding how quantum effects might play a role in the neurophysiological processes underlying cognition and even consciousness (see Atmanspacher 2015; Pylkkänen 2018). From the perspective of the ontological interpretation the important question is whether there are some "quantum sites" in the brain where the quantum potential can have a non-negligible effect. Hiley and Pylkkänen (2005) discuss one such possibility by applying the quantum potential approach to Beck and Eccles's (1992) ideas about the role of quantum mechanics in synaptic exocytosis. Beck and Eccles suggested that the appearance of low transition probabilities in synaptic exocytosis implies that there exists an activation barrier against the opening of an ion channel in the presynaptic vesicular grid. Hiley and Pylkkänen (2005: 21–2) suggested that it is the action of the quantum potential that effectively reduces the height of the potential barrier to increase the probability of exocytosis (this is an example of quantum tunneling). In the extended ontological interpretation the higher order "mental" fields can then in a natural way influence the quantum potential and in this way control synaptic exocytosis. Quantum tunneling in synaptic communication has recently been discussed in detail by Danko Georgiev (2018) and I suggest that a promising way to develop Bohmian quantum brain theory further is to examine Georgiev's proposals in the light of the richer picture of quantum processes that the ontological interpretation provides. Another possibility is to apply the Bohm scheme to recent proposals that quantum coherence is involved in ion channel conductivity and selectivity (Salari et al. 2017). By controlling the quantum potential in the ion channels, the higher-order mental fields might then be able to control the triggering of action potentials. Such speculations require a careful consideration of the problem of decoherence.

Alternatively, one might look in the light of the ontological interpretation at Penrose and Hameroff's proposal that consciousness depends on biologically "orchestrated" coherent quantum processes in collections of microtubules within brain neurons. Penrose and Hameroff (unlike Bohm and Hiley) give high importance to the role of the orchestrated objective reduction (collapse) of the quantum state in the regulation of axonal firings and the control of conscious behavior (see Hameroff and Penrose 2014). In the ontological interpretation such regulation would take place through the higher-order "mental" fields orchestrating the quantum potential (which need not involve collapse).

Let us summarize Bohmian mental causation. The Bohm-Hiley scheme gives a key causal role to information at the quantum level, by suggesting that information about the environment of the quantum particles is encoded in the wave function and guides the particles. It is reasonable to postulate a *hierarchy of fields* of information in complex systems such as brains. If free will and spontaneity is possible at the higher levels of information, then the hierarchy enables free will to guide physical action. The principle of the causal closure of the physical domain needs to be abandoned or else modified to include (traditionally non-physical) causal features such as information (for a more thorough discussion, see Pylkkänen 2007, 2017).

14.5 Conclusion: Two Kinds of Mental Causation, Contra the Received View of Physicalism

In this paper we first considered how causal anti-fundamentalism suggests a possible route to mental causation. According to causal anti-fundamentalism the physical world is not governed by any fundamental relation of cause and effect. But it typically assumes that fundamental physical facts ground causal facts, including those involving consciousness. Conscious experiences can then be seen as local events which determine events in their future.

We also saw that there are reasons to think that causal anti-fundamentalism is incorrect. But we noted that there is another way to challenge the received view, namely by making use of the ontological interpretation of quantum theory. This interpretation suggests that information plays a fundamental causal role in the physical world. We argued that it is reasonable to assume that there exists a hierarchy of levels of information, with two-way causal influences between levels. Mental/conscious states are part of this hierarchy and can thus causally affect and be affected by physical processes at lower levels of the hierarchy.

In this short paper our discussion has been schematic, and many of the issues need to be explored more carefully in later research. For example, how to reconcile the suggestion that causality can be eliminated in classical physics, with the suggestion that it cannot be eliminated in (usual) quantum mechanics (and possibly in the Bohm-Hiley theory)? Further, the Bohm-Hiley theory is non-local, so what is the role of causality in those situations where a non-local quantum potential has a non-negligible effect (e.g. an EPR-type experiment; cf. Fenton-Glynn and Kroedel 2015; Walleczek 2016; Musser 2015)?

Acknowledgements Earlier versions of this paper have been presented at the Annual Colloqium of the Philosophical Society of Finland in Helsinki, January 2018; at The Science of Consciousness conference in Tucson, USA, April 2018; at the Parmenides Foundation workshop "Rethinking Matter, Life and Mind", Tegernsee, Germany, September 2018; and at the "Scientific metaphysics" workshop at the University of Helsinki, January 2019. I thank the participants of these events for valuable comments and questions. The work for this paper was partially funded by the Fetzer Franklin Fund of the John E. Fetzer Memorial Trust.

References

Andersen, F., Anjum, R. L., & Mumford, S. (2018). Causation and quantum mechanics. In R. L. Anjum & S. Mumford (Eds.), *What tends to be. The philosophy of dispositional modality*. London: Routledge.

Atmanspacher, H. (2015). *Quantum approaches to consciousness*, 1 (Summer 2015 ed.), E. N. Zalta (Ed.). http://plato.stanford.edu/archives/sum2015/entries/qt-consciousness/

Beck, F., & Eccles, J. (1992). Quantum aspects of brain activity and the role of consciousness. *Proceedings of the National Academy of Sciences, 89*(23), 11357–11361.
Blanchard, T. (2016). Physics and causation. *Philosophy Compass, 11*, 256–266.
Bliss, R., & Trogdon, K. (2016). Metaphysical grounding. *The Stanford Encyclopedia of Philosophy* (Winter 2016 ed.), E. N. Zalta (Ed.). https://plato.stanford.edu/archives/win2016/entries/grounding/
Bohm, D. (1951). *Quantum theory*. Englewood Cliffs: Prentice-Hall. Dover edition 1989.
Bohm, D. (1957/1984). *Causality and chance in modern physics*. London: Routledge.
Bohm, D. (1990). A new theory of the relationship of mind and matter. *Philosophical Psychology, 3*, 271–286.
Bohm, D., & Hiley, B. J. (1987). An ontological basis for quantum theory: I. Non-relativistic particle systems. *Physics Reports, 144*, 323–348.
Bohm, D., & Hiley, B. J. (1993). *The undivided universe: An ontological interpretation of quantum theory*. London: Routledge.
Cartwright, N. (1979). Causal laws and effective strategies. *Nous, 13*, 419–437.
Fenton-Glynn, L. & Kroedel, T. (2015). Relativity, quantum entanglement, counterfactuals and causation. *British Journal for the Philosophy of Science, 66*(1), 45–67.
Frisch, M. (2012). No place for causes? Causal skepticism in physics. *European Journal for Philosophy of Science, 2*, 331–336.
Georgiev, D. (2018). *Quantum information and consciousness: A gentle introduction*. Boca Raton: CRC Press.
Goff, P., Seager, W., & Allen-Hermanson, S. (2017). Panpsychism, *The Stanford Encyclopedia of Philosophy* (Winter 2017 ed.), E. N. Zalta (Ed.) https://plato.stanford.edu/archives/win2017/entries/panpsychism/
Goldstein, S. (2013) Bohmian mechanics. *The Stanford Encyclopedia of Philosophy*, E. N. Zalta (Ed.). http://plato.stanford.edu/archives/spr2013/entries/qm-bohm/
Hameroff, & Penrose. (2014). Consciousness in the universe: A review of the 'Orch Or' theory. *Physics of Life Reviews, 11*(2014), 39–78.
Hiley, B. J., & Pylkkänen, P. (2005). Can mind affect matter via active information? *Mind and Matter, 3*(2), 7–26. http://www.mindmatter.de/resources/pdf/hileywww.pdf
McGinn, C. (2006). Hard questions. *Journal of Consciousness Studies, 13*(10–11), 90–99.
Musser, G. (2015). *Spooky action at a distance*. New York: Farrar, Straus and Giroux.
Price, H., & Corry, R. (Eds.). (2007). *Causation, physics, and the constitution of reality: Russell's republic revisited*. Oxford: Clarendon Press.
Pylkkänen, P. (2007). *Mind, matter and the implicate order*. Heidelberg/New York: Springer.
Pylkkänen, P. (2017). Is there room in quantum ontology for a genuine causal role of consciousness? In A. Khrennikov & E. Haven (Eds.), *The Palgrave handbook of quantum models in social science*. London: Palgrave Macmillan.
Pylkkänen, P. (2018). Quantum theories of consciousness. In R. Gennaro (Ed.), *The Routledge companion to consciousness*. London: Routledge.
Pylkkänen, P. (forthcoming). A quantum cure for panphobia. In W. Seager (Ed.), *Routledge handbook of panpsychism*. London: Routledge.
Robb, D., & Heil, J. (2018). Mental causation. *The Stanford Encyclopedia of Philosophy*, E. N. Zalta, Ed. (Winter 2018 ed.). https://plato.stanford.edu/archives/win2018/entries/mental-causation/
Russell, B. (1913). On the notion of cause. *Proceedings of the Aristotelian Society, 13*, 1–26.
Salari, V., Naeij, H. & Shafiee, A. (2017). Quantum interference and selectivity through biological ion channels. *Scientific Reports 7*, 41625.
Strawson, G. (2006). Realistic monism – Why physicalism entails panpsychism. *Journal of Consciousness Studies, 13*(10–11), 3–31.
Sundström, P., & Vassen, B. (2017). Description of the "Where is there causation?" -workshop, Umeå University, 27–29 October 2017, see https://philevents.org/event/show/34954
Walleczek, J. (2016). The super-indeterminism in orthodox quantum mechanics does not implicate the reality of experimenter free will. *Journal of Physics: Conference Series, 701*, 012005.

Chapter 15
Introspection and Superposition

Paul Skokowski

Abstract An Everettian interpretation of quantum mechanics given by David Albert claims that a competent observer of a superposition would be deceived when introspecting her own perceptual beliefs. A careful accounting of the belief states of the observer, together with an understanding of the linearity of operators that represent observables in quantum mechanics, shows that this claim is mistaken. A competent observer's introspection about her perceptual belief of the measurement of a superposition cannot be a deception.

15.1 Introduction

In *Quantum Mechanics and Experience*, David Albert claims that the linearity of operators that represent observables in quantum mechanics leads to cases where a single eigenvalue can be elicited from a superpositional state, yielding potentially puzzling results. Though this is true due to the mathematical properties of linearity, there are problems with the two examples Albert chooses to illustrate what is puzzling about these properties. Understanding the problem with the first, simple, example that Albert gives of a particle in a box helps us to understand the deeper and more interesting problem of the second example: the nature of an observer's mental states when she is observing a superposition. In both cases it turns out the eigenvalues necessary for obtaining the results Albert claims for the superpositions

---

A huge thanks to my colleagues Reed Guy, Harvey Brown, and John Perry for helping me struggle through some of these concepts. All mistakes and misunderstandings are entirely my own, however. I would also like to thank attendees at the 2018 International Conference on Quanta and Mind for helpful comments and suggestions. Also, special thanks are due to Christopher Skokowski for advice on using LaTeX.

P. Skokowski (✉)
St Edmund Hall, Oxford University, Queen's Lane, Oxford, UK
e-mail: paul.skokowski@seh.ox.ac.uk

J. Acacio de Barros, C. Montemayor (eds.), *Quanta and Mind*, Synthese Library 414, https://doi.org/10.1007/978-3-030-21908-6_15

in question require additional eigenstates, and additional operators that are specific to those additional eigenstates. This analysis raises the question of whether an observer of a superposition is radically deceived in the way Albert claims. Indeed, we will see that by carefully accounting for the observer's brain states – and in particular her belief states – throughout the process of her observing a spin-measurement, and by assigning the appropriate eigenstates and eigenvalues to those states, that the observer is not deceived in the manner claimed by Albert.

15.2 Observing a Superposition

In Chapter 6 of *Quantum Mechanics and Experience*, David Albert considers a human experimenter he calls "h" – let's call her Hilda – who measures the spin of an electron. In this paper, I will use Albert's notation to follow his development of the quantum mechanical states during the experiment. In particular, Albert uses his own notation to describe orthogonal electron spin states. He calls one axis the "hardness" axis, and the other, orthogonal, axis the "color" axis. Albert designates the spin directions "+" and "−" along the hardness axis as "hard" and "soft", and he designates the spin directions "+" and "−" along the color axis as "black" and "white." For the purposes of this paper, let's designate the hardness axis as the x-axis, and the color axis the z-axis. This will mean that spin-up on the x-axis, represented by the ket vector $|+x\rangle$, will be given by the ket vector $|hard\rangle$, and spin-down along the x-axis, represented by the ket vector $|-x\rangle$, will be given by the ket vector $|soft\rangle$. Similarly, spin-up on the z-axis, represented by the ket vector $|+z\rangle$, will be given by the ket vector $|black\rangle$, and spin-down along the x-axis, represented by the ket vector $|-z\rangle$, will be given by the ket vector $|white\rangle$.

In the experiment Albert describes, Hilda measures the color of a hard electron. Since a hard electron is represented by the ket vector $|+x\rangle$, this vector can be expanded as a superposition of states in the z-direction:

$$|+x\rangle = \frac{1}{\sqrt{2}}|+z\rangle + \frac{1}{\sqrt{2}}|-z\rangle$$

or, in Albert's notation:

$$|\text{hard}\rangle = \frac{1}{\sqrt{2}}|\text{black}\rangle + \frac{1}{\sqrt{2}}|\text{white}\rangle$$

Now, when the experiment begins, a hard electron is headed to a color spin measuring device, and Hilda is watching the device. At this stage, before the electron enters the measuring device, the device is in its "ready" state (ready to make a spin measurement), and Hilda is also in her "ready" state (ready to observe the reading on the measuring device.) Albert uses the subscripts "*h*" for Hilda, "*m*" for the measuring device, and "*e*" for the electron. At this stage, the system is in the state:

$$|\text{ready}\rangle_h|\text{ready}\rangle_m|\text{hard}\rangle_e$$

When the experiment is over, the above state evolves to a superposition of states, one component of which has a black electron emerging, the measuring device detecting "black" and Hilda believing the detector reads "black", and the other component of which has a white electron emerging, the measuring device detecting "white" and Hilda believing the detector reads "white" (Albert 1992, p. 112). This result is ensured from the linear dynamical equations of motion (Albert 74, 75), and from the stipulation by Albert that Hilda "is a competent observer of the positions of pointers"(Albert 77).

Albert writes out this superpositional state of Hilda (h), the measuring device (m) and the measured electron (e) in equation (6.1) as follows:

$$\frac{1}{\sqrt{2}}\big(|\text{believes } e \text{ black}\rangle_h|\text{"black"}\rangle_m|\text{black}\rangle_e \quad (6.1)$$
$$+\ |\text{believes } e \text{ white}\rangle_h|\text{"white"}\rangle_m|\text{white}\rangle_e\big)$$

For the remainder of this paper, I will shorten this notation a bit by writing 'black' as 'b' and 'white' as 'w'. We then get for Albert's equation (6.1) the following:

$$\frac{1}{\sqrt{2}}\big(|\text{believes } e \text{ b}\rangle_h|\text{"b"}\rangle_m|\text{b}\rangle_e + |\text{believes } e \text{ w}\rangle_h|\text{"w"}\rangle_m|\text{w}\rangle_e\big) \quad (6.1)$$

Albert now considers the interesting question of what it is like to be in the superposition of (6.1). That is, he considers the case of the linear quantum mechanical equations of motion being "the true and complete equations of motion of the whole world"; that is, an Everettian interpretation (Albert, p. 113).

Albert considers asking Hilda about her present belief about the color of the electron which has just been measured. But that's a problem, because Hilda appears from (6.1) to be in a superposition of belief states: one is a belief that the electron is black, and the other is a belief that the electron is white. And so the state of the world after asking that question will be a superposition of Hilda saying "black" and of her saying "white".

So, Albert suggests asking a different question, which is for Hilda to answer whether she has "...any definite particular belief ...about the value of the color of this electron" (Albert, 118). According to Albert, in either component of the superposition, Hilda will answer "yes". And so, considering the observable "Do you have any definite particular belief about the color of this electron?" the eigenvalue will be "yes" for both components of the superposition. This means, "by the linearity of operators that represent observables of quantum mechanical systems" (Albert, 117) that this observable applied to the superposition will also yield the eigenvalue, and hence the answer, "yes".

As Albert says, saying "yes" is an "observable property" of Hilda, "...and consequently (in particular) it will be an observable property of (6.1)" (Albert, 118).

15.3 Positions and Boxes

Albert introduces this 'linearity of operators' in preparation for the Hilda case on the previous page, where he discusses the linearity properties of operators for observables. Albert says:

> Note, to begin with, that it follows from the linearity of the operators that represent observables of quantum-mechanical systems ... that if any observable O of any quantum-mechanical system S has some particular determinate value in the state $|A\rangle_S$, and if O also has that same determinate value in some other state $|B\rangle_S$, then O will necessarily also have precisely that same determinate value in any linear superposition of those two states. (Albert, 117)

What Albert is describing here, of course, is an operator O which has the same eigenvalue, let's call it λ, when operating on either of the two states $|A\rangle_S$ and $|B\rangle_S$. So if we consider O operating on a linear superposition of these two states, such as $|A\rangle_S + |B\rangle_S$, then we would have:

$$O(a|A\rangle_S + b|B\rangle_S) = aO|A\rangle_S + bO|B\rangle_S = a\lambda|A\rangle_S + b\lambda|B\rangle_S = \lambda(a|A\rangle_S + b|B\rangle_S)$$

This is a property of linear operators which happen to yield the same eigenvalue for two different eigenvectors in a superposition. So, for an operator with this property for the two eigenstates $|A\rangle_S$ and $|B\rangle_S$, a superposition of these two states is also an eigenstate with the same eigenvalue, λ.

In a footnote directly below the above quote, Albert elaborates:

> That's an entirely commonsensical way for observables to behave, if you think it through. Suppose, for example, that there's a particle which is in a superposition of being located in the right half and in the left half of a certain box. What the linearity of the observables of a particle like that is going to entail (or rather, one of the things that it's going to entail) is that particle is in an eigenstate of the observable "is the particle anywhere in the box at all?" with eigenvalue "yes." (Albert, 117)

At first glance, this makes sense. But there's a slight problem, as I see it. And the problem is that it is difficult to construct an operator which has the property that Albert is highlighting here. Let me give an example. I will borrow the notation Albert uses on p. 46 of the same book.

Consider the superpositional state $|\Phi\rangle = \frac{1}{\sqrt{2}}|X = L\rangle + \frac{1}{\sqrt{2}}|X = R\rangle$. The eigenstate $|X = L\rangle$, when operated on by the position operator X, yields the eigenvalue L. The eigenstate $|X = R\rangle$, when operated on by the position operator X, yields the eigenvalue R. The superpositional state is of course a (very) simplistic representation of finding a particle with equal probability either in the left side of a box (at point $x = L$), or in the right side of a box (at point $x = R$), with the value of zero at all other points.

The problem here seems to be that this system has one degree of freedom, x, and that as such, the only operator that can operate on it is the position operator X. But when operating on this system with operator X, the first eigenstate $|X = L\rangle$, can yield only the eigenvalue L, and the second eigenstate $|X = R\rangle$, can only yield the eigenvalue R. And it is apparent that $L \neq R$. So these two eigenvectors can not in fact have an eigenvalue in common, which means that the superpositional state also cannot have an eigenvalue in common with the two states $|X = L\rangle$ and $|X = R\rangle$ which compose it.

Consider an example given by Albert in chapters 1 and 2 of *Quantum Mechanics and Experience*. Here, Albert describes a two-path experiment, which results in a superposition of states for the electron involved (Albert, 11 and 55). In this experiment a white electron is sent into a color box (a Stern-Gerlach device), and Albert explains how the electron ends up in a superposition of states, where the superposition includes a component where the electron takes one path, which he calls path 'h' (where the electron emerges as a 'hard' electron) and a component where the electron takes the other path, which he calls path 's' (where the electron emerges as a 'soft' electron). Albert uses the following language when discussing the two-path experiment: "Electrons ... do not take route h and do not take route s and do not take both of these routes and do not take neither of those routes..." (Albert, 11 and 55) We can apply the same language to the superposition of position states $|\Phi\rangle$ encountered in our example above. Adapting this language to $|\Phi\rangle$ would yield: "The particle is not in the left side of the box and not in the right side of the box and is not in both sides of the box and not in neither side of the box." And if this is true of $|\Phi\rangle$, then how can we assure ourselves that "the particle is in an eigenstate of the observable "is the particle anywhere in the box at all?" with eigenvalue "yes"" without finding a common eigenvalue for the two states $|X = L\rangle$ and $|X = R\rangle$ composing the superposition? The language just given us by Albert for a superposition of states appears to deny us this assurance.

Again, the problem is that the only operator that can yield position information for the particle in the box is the position operator itself: X. This operator does not seem to be of the form "Is the particle anywhere in the box at all", but rather of a much simpler form, "the position of the particle is x".

Therefore I'm skeptical that Albert can be granted the claim that there is an operator of the form "Is the particle anywhere in the box at all", which yields the eigenvalue "yes", for this superposition of state vectors.

Let me however propose a solution; one that solves Albert's problem of the particle in the box, but in so doing shows that the original setup for the problem was slightly disingenuous. As Jeff Barrett points out "Each physical quantity that one might observe corresponds to a Hermitian operator on an appropriate Hilbert space"(Barrett 1999, 32). The key word here is 'appropriate.' There are two properties that are involved when considering the particle described above. One is the actual position of the particle, which is being represented by the state vector $|\Phi\rangle$. The other is whether the particle is in a box. We see first, that the position can be obtained by applying the position operator X. The eigenstates for the X operator look like $|X = L\rangle$ and $|X = R\rangle$. Second, whether the particle is in the box or not can be found by applying what I'll suggestively call the 'Box' operator: *Box*. The eigenstates for the Box operator look like $|Inside\ the\ Box = \lambda\rangle$ and $|Inside\ the\ Box = \gamma\rangle$, where λ = "Yes" and γ = "No."

Of course, now our wavefunction is going to be different, so let's call it $|\Phi'\rangle$. Since Albert has told us in advance that the possible values for the particle will be inside the box, we can construct the new wavefunction $|\Phi'\rangle$:

$$|\Phi'\rangle = \frac{1}{\sqrt{2}}|X = L\rangle|Inside\ the\ Box = \lambda\rangle + \frac{1}{\sqrt{2}}|X = R\rangle|Inside\ the\ Box = \lambda\rangle$$

For simplicity we have assumed the odds of finding the particle in either side of the box are equal.

Now we can operate on $|\Phi'\rangle$ with the Box operator to get Albert's result:

$$\begin{aligned} Box|\Phi'\rangle = \ & Box\Big(\frac{1}{\sqrt{2}}|X = L\rangle\ |Inside\ the\ Box = \lambda\rangle \\ & + \frac{1}{\sqrt{2}}|X = R\rangle\ |Inside\ the\ Box = \lambda\rangle\Big) \\ = \ & \frac{1}{\sqrt{2}}|X = L\rangle\ Box|Inside\ the\ Box = \lambda\rangle \\ & + \frac{1}{\sqrt{2}}|X = R\rangle\ Box|Inside\ the\ Box = \lambda\rangle \\ = \ & \frac{1}{\sqrt{2}}|X = L\rangle\ \lambda|Inside\ the\ Box = \lambda\rangle \\ & + \frac{1}{\sqrt{2}}|X = R\rangle\ \lambda|Inside\ the\ Box = \lambda\rangle \\ = \ & \lambda\Big(\frac{1}{\sqrt{2}}|X = L\rangle|Inside\ the\ Box = \lambda\rangle \\ & + \frac{1}{\sqrt{2}}|X = R\rangle|Inside\ the\ Box = \lambda\rangle\Big) \\ Box|\Phi'\rangle = \ & \lambda|\Phi'\rangle \end{aligned}$$

In other words, it turns out we needed an additional eigenfunction in order to achieve the result Albert requires, that is, answering a question of whether or not the particle is in the box.

15.4 Introspection and Superposition

Now let us return to Albert's original example, where he asks Hilda whether she has "…any definite particular belief …about the value of the color of this electron." (Albert, 118)

What Albert is asking Hilda to report is an introspective belief, which is a belief about a belief. When we ask someone to report on some belief they are holding, they need to introspect in order to access and report that belief. Suppose Hilda sees a green patch on a grey screen in front of her. Then we would say that she has the occurrent belief that the patch is green. Indeed, this occurrent belief would involve visual cortex (Zeki 1993; Seymour et al. 2016; Lee et al. 1998). In such a case, upon being asked what the color of the patch is in front of her, Hilda reports that "the patch is green." Now let's ask Hilda to report if she has a definite belief about the color of a patch in front of her. Hilda might ask in this case, "are you asking me what the color of the patch is?", to which we would answer, "No, we are asking you if you now have a belief about the color of the patch in front of you. In particular, we are asking you to introspect your belief about the color of the patch in order to verify that you have a definite belief about the color of the patch." In this case, Hilda will form an introspective belief – a belief about a belief – because she will need to check her current beliefs to verify that she has a belief about a colored patch in front of her. This is a belief that will have another belief as a content; and in particular, the content of this introspective belief will be Hilda's occurrent belief.

Note that the fine-grainedness of mental contents ensures that Hilda's belief, or introspective state, or sensation that the electron is white will be different from her belief, or introspective state, or sensation that the electron is black (Tye 1995; Dretske 1995; Perry 1977; Frege 1892). This fine-grainedness is important: the intentional content of a belief about whiteness is different from the intentional content of a belief about blackness.[1] Such beliefs then, will always differ. So, any sort of mental state with a content about whiteness will differ from any mental state about blackness. And of course, it is the content of the state that allows us to call such states mental representations in the first place. The content of a mental state also helps distinguish that type of state from other states, and in a causal theory of mental content, such states will be distinguished because of their differences in content, vehicle, and causal role (Skokowski 1999, 2018).

In order to have an introspective belief that can be used to report the occurrent belief being introspected, this introspective belief must be causally and physically connected with the occurrent perceptual belief through neural connections and neural firings. So, Hilda's introspective belief about a given perceptual belief will involve specific neural connections – as part of the introspective belief vehicle – to the specific portion of visual cortex which is involved with the exemplification of her perceptual belief which itself involves a different set of neurons. In addition, when Hilda has this introspective belief, then specific neural firings of the sort associated with the particular connections to this region of visual cortex will occur. That is, the action potentials that occur within the introspective belief in virtue of its connection with the color belief will be specific to that particular connection between the two

[1] Note that this difference in content holds whether the content is the position of a pointer towards either 'black' or 'white' or whether the content is actual color content 'black' or 'white.' The intentional content will be fine-grained in either case.

sets of neurons. This introspective belief vehicle will therefore be comprised of a specific set of neurons, which exemplify a certain pattern of neural firing (action potentials), when it occurs in Hilda's brain. Indeed, imaging studies appear to show that introspective beliefs occur in pre-frontal cortex (Fleming et al. 2010).

And this analysis will apply for any of Hilda's mental states that are tasked with representing the content of her occurrent perceptual belief. Any such mental state will need to be causally connected to the occurrent perceptual belief, and the intentional content of the analyzing mental state will be fine-grained and hence about that occurrent perceptual belief and its content. So: if Hilda has a mental state M which is about her occurrent perceptual state and its content – where M is an introspective or some other self-analyzing state in Albert's sense – then this mental state M will have a fine-grained content that is about that occurrent perceptual belief and its content. Hence any query to Hilda about an occurrent perceptual belief state of hers will depend for its answer on a mental state of hers which represents that occurrent perceptual belief, and the intentional content of that representing mental state will have a fine-grainedness that tracks the fine-grainedness of the occurrent perceptual belief it represents.[2]

So for example, the eigenstate corresponding to Hilda's introspection of her belief that the electron is white will contain her introspective state in pre-frontal cortex as well as her perceptual belief in visual cortex, and would therefore be written:

$$|\text{introspect } e \text{ w}\rangle_{h-PF}|\text{believes } e \text{ w}\rangle_{h-VC}|\text{“w”}\rangle_m|\text{w}\rangle_e$$

where the subscripts "*h-PF*" stand for the introspective state in Hilda's pre-frontal cortex and the subscripts "*h-VC*" stand for the perceptual belief involving Hilda's visual cortex.

Let's apply Albert's analysis to how the initial state of Hilda's brain, the measuring device she is observing, and the electron evolves from being "ready" to how things end up after the measurement. Before the measurement we have:

$$|\text{PF ready}\rangle_{h-PF}|\text{VC ready}\rangle_{h-VC}|\text{ready}\rangle_m\left(\frac{1}{\sqrt{2}}|\text{b}\rangle_e + \frac{1}{\sqrt{2}}|\text{w}\rangle_e\right)$$

where the first state is the 'ready' state of the pre-frontal cortex, the second state is the 'ready' state of the visual cortex, the third state is the 'ready' state of the measuring device, and the final state is the superposition of the spin directions black and white (that is, a 'hard' electron written in the black and white basis) for the electron that is about to be passed through the detector.

After the measurement, this state evolves into a superposition, call it $|\Psi\rangle$, of the form:

[2]If the representing state did not have this fine-grainedness, then Hilda would not be capable of answering queries about the content of the perceptual state in question.

$$|\Psi\rangle = \frac{1}{\sqrt{2}}\big(|\text{introspect } e \text{ b}\rangle_{h-PF}|\text{believes } e \text{ b}\rangle_{h-VC}|\text{“b”}\rangle_m|\text{b}\rangle_e$$
$$+ \ |\text{introspect } e \text{ w}\rangle_{h-PF}|\text{believes } e \text{ w}\rangle_{h-VC}|\text{“w”}\rangle_m|\text{w}\rangle_e\big)$$

Note that the first two eigenstates of both terms in the superposition represent different mental/belief states of Hilda's. As such, they will have different neural vehicles and contents from one another. In addition, were we to measure them, they would yield different eigenvalues. In the first term of the superposition, the first (introspective) state would give a measurement of a state of an introspection of a belief that the electron is black, and the second state would give a measurement of a perceptual belief that the electron is black. In the second term of the superposition, the first (introspective) state would give a measurement of a state of an introspection of a belief that the electron is white, and the second state would give a measurement of a perceptual belief that the electron is white. All these states differ in vehicle (the particular state in cortex) and content, which has to be the case given the fine-grainedness of mental states.

Since each component of this superposition yields a different eigenvalue for its corresponding operator, then there will not be a common eigenvalue between the two components which would serve as an eigenvalue for an operator on the superposition taken as a whole. This is because of the fine-grainedness of mental states. Since each mental state in the superposition has its own vehicle and fine-grained intentional content, then each one will differ: each one will have a different eigenvalue from the other. This has to be the case for mental states, if they are to be individuated qua mental states. An introspective state about a belief that an electron is white is different from an introspective state about a belief that an electron is black. A perceptual belief that an electron is white is different from a perceptual belief that an electron is white. The vehicles differ: each state – introspective or perceptual – is comprised of different neurons and action potentials, so the vehicles are exemplified as different physical states.

There are two important things to note here. First, the state of the system corresponding to Hilda's introspection of her belief that the electron is white will not be given by equation (6.1) as Albert initially claimed. Instead, the state of the system will contain her introspective state in pre-frontal cortex as well as her perceptual belief involving visual cortex, as spelled out above with state $|\Psi\rangle$. Second, the superpositional state $|\Psi\rangle$ as it stands does not yield the result that Albert has claimed, which is that there is a measurable eigenvalue that is common to both components of the superposition. If there were such an eigenvalue, then there would be an observable property of Hilda for the superposition of brain states she finds herself in according to the linear equations of motion.

Albert's solution to this problem is to have Hilda answer "yes" to a question about her mental state. Again, the question is for Hilda "tell me ... whether or not you now have any particular definite belief ... about the value of the color of this electron"(Albert, 118).

But note that there is a peculiarity about answering such a question. Answers to questions are formulated in a different part of the brain: Broca's area. And Broca's area is an area that is tasked with linguistic output – not with introspection. These linguistic outputs include unconscious grammatical processing and the signals required to form the mouth and tongue in a particular configuration, exhaling breath in a certain manner, opening and closing the nasal passages, and so forth.[3] Introspection is tasked with producing beliefs about beliefs (Dretske 1995), whereas Broca's area is not: it is tasked with producing linguistic outputs (Pinker 1994, 1997).

The outputs of intentional mental states are not to be confused with mental states themselves. Consider the simple example of taking a drink. I believe there's a bottle of water in front of me and I desire a drink. These mental states cause me to reach for the bottle and bring it to my lips. The belief and desire are mental states with intentional contents, but the reaching and lifting are outputs which are caused by these representational states. These outputs are not themselves representational states: that is, states with intentional contents from a function to represent properties in the environment, and executive capacities to cause action (Dretske 1988; Skokowski 1999). They are instead outputs of representational states: causal consequences of the (mental) states which have the executive capacities and intentional contents.

Albert's question therefore, is designed to detect a common measurable property in the Hilda-spin-detector-system by detecting a common output, and not by detecting a common property of the introspective representational eigenstates themselves. This is peculiar, because what is presumably at issue is the content of Hilda's own introspective and occurrent perceptual beliefs. Indeed, Albert says that "when a state like (6.1) obtains, (Hilda) is apparently going to be radically deceived even about what her own occurrent mental state is" (Albert 118; my parentheses). The immediate problem here is that deception is surely a mental state. Hilda's being deceived is going to amount to Hilda having a certain kind of mental state: a mental state that is false. This is indeed a crucial property of beliefs, what Brentano called "the mark of the mental": beliefs are intentional states with the capacity to be false (Brentano 1874; Dretske 1988). But Hilda's representational states – her beliefs – are not being measured here, even though as physical states they are in principle measurable.

But again, Albert has precluded this latter option by prescribing to Hilda, "Don't tell me whether you believe the electron to be black or you believe it to be white..." The reason for this prescription is clear: The eigenvalue for asking Hilda "Do you introspect you are perceiving black?" will be different on both sides of the superposition, as it will if we decide instead to ask Hilda "Do you introspect you are perceiving white?", and so there will be no common eigenvalue for the superposition $|\Psi\rangle$ if either of these questions is asked. So Albert must ask a different question. But that means that, rather than measuring Hilda's introspective states and their

[3]Because introspection is conscious (Dretske 1995; Moore 1903), and linguistic processing is unconscious (Pinker 1994), the latter is not a candidate for introspective beliefs.

contents with an operator operating on them directly, we are instead being asked to measure a common output of those introspections. The focus has been shifted from introspection itself to a common output of introspections.

The problem of shifting the focus in this way can be illustrated by an example. Note that if we can detect a common eigenvalue in the Hilda-spin-detector superposition by means of a common output, rather than by the representational state itself, then we should be able to detect a common eigenvalue in a measuring device by a common output as well. After all, what is important about an electron-spin measuring device is the representational state it ends up in: pointing to white, or pointing to black. These pointer states are representational states with a content: they are *about* something, and crucially, they are about whether the measured electron is black or white. That's what makes them representational in a way appropriate for measuring the actual spin of a particular electron. But if there's a common output for these two states, then, according to Albert, we can use this common output to show that the device is, using Albert's own description, "radically deceived even about what its own occurrent [representational] state is."

Suppose, for example, that the needle on our electron-spin measuring device slides on the x-axis. It slides left for a white electron and right for a black electron. Now suppose we notice that whenever the measuring device registers the spin of an electron, there is a slight vibration in the y-axis normal to the x-axis. This vibration is the same whether the electron is black or white. So we place a vibration detector on the y-axis that detects when a vibration occurs.

Before we had this vibration detector attached, our superposition looked like:

$$|\Phi\rangle = \frac{1}{\sqrt{2}}\big(|\text{``black''}\rangle_m|\text{black}\rangle_e + |\text{``white''}\rangle_m|\text{white}\rangle_e\big)$$

And after we attached the vibration detector we have:

$$|\Phi'\rangle = \frac{1}{\sqrt{2}}\big(|\text{``y − vibration''}\rangle_v|\text{``black''}\rangle_m|\text{black}\rangle_e + |\text{``y − vibration''}\rangle_v|\text{``white''}\rangle_m|\text{white}\rangle_e\big)$$

Call the magnitude of this vibration λ. Then we can see, by linearity, that a measurement of this vibration by an operator O is an observable property of the superpositional state as well as of each component of the superposition. So, by putting a hard electron through a color detector, we can measure the observable property λ:

$$O|\Phi'\rangle = \lambda|\Phi'\rangle$$

But notice that measuring a vibration like this – that is, an output of a representational/detecting state – does not mean that the detector is radically deceived about what it's detecting. It just means that the detector produces a measurable vibration regardless of whether it has collapsed to one component of the superposition or the other, or that, if Everett is correct, this observable will be measurable even in a superposition (by virtue of linearity). A measurement of a y-vibration makes no claim about the final position of the pointer in the x-direction and hence the spin of the electron. This is explicitly spelled out in the state vector $|\Phi'\rangle$, where the two properties occupy different eigenstates, for example, $|\text{"y} - \text{vibration"}\rangle_v$ and $|\text{"black"}\rangle_m$, that yield different physical properties for the device. Measurement of one property of the device does not mean deception about another property, as these are separate eigenstates with their own associated operators and eigenvalues. So the detector is not deceived. There is just a common measurable output which occurs regardless of the QM interpretation.

And now we can say the same about the Hilda case. When we consider the linguistic output from Broca's area, then before the measurement we have:

$$|\text{B ready}\rangle_{h-B}|\text{PF ready}\rangle_{h-PF}|\text{VC ready}\rangle_{h-VC}|\text{ready}\rangle_m\big(\frac{1}{\sqrt{2}}|\text{b}\rangle_e + \frac{1}{\sqrt{2}}|\text{w}\rangle_e\big)$$

where the first state is the 'ready' state of Hilda's Broca's Area, and the rest of the 'ready' states are defined as before: Hilda's pre-frontal cortex, her visual cortex, the 'ready' state of the measuring device, and the final state is the superposition of black and white spin for the electron that is about to be passed through the detector. Once again, Hilda's eigenstates are designated by subscripts *'h-B'*, *'h-PF'* and *'h-VC'*.

After the measurement, this state evolves into a superposition of the form $|\Psi'\rangle$, which is different from the superposition $|\Psi\rangle$ we considered earlier (and which was itself different from Eqn. (6.1) given by Albert):

$$|\Psi'\rangle = \frac{1}{\sqrt{2}}\big(|\text{"Yes"}\rangle_{h-B}|\text{introspect } e \text{ b}\rangle_{h-PF}|\text{believes } e \text{ b}\rangle_{h-VC}|\text{"b"}\rangle_m|\text{b}\rangle_e$$
$$+ \ |\text{"Yes"}\rangle_{h-B}|\text{introspect } e \text{ w}\rangle_{h-PF}|\text{believes } e \text{ w}\rangle_{h-VC}|\text{"w"}\rangle_m|\text{w}\rangle_e\big)$$

Then we can see, by linearity, that a measurement of Hilda's linguistic output will be an observable property of the superpositional state as well as of each component of the superposition. That is, when the system $|\Psi'\rangle$, which includes Hilda's states, is asked whether Hilda has some definite belief in the way prescribed by Albert, where this question is the operator O, then she will answer "Yes":

$$O|\Psi'\rangle = \text{"Yes"}|\Psi'\rangle$$

And this result is by virtue of this operator O operating on Hilda's eigenstate $|\text{"Yes"}\rangle_{h-B}$.

But this is like the detector example given immediately above. 'Measuring' an answer like this – that is, an output of an introspective/representational state – does not mean that Hilda is radically deceived about what she's introspecting. It just means that Hilda produces the answer "Yes" regardless of whether she has collapsed to one component of the superposition or the other, or that, if Everett is correct, this answer will be measurable even in a superposition (by virtue of linearity). A measurement of a spoken output "Yes" makes no claim about Hilda's occurrent introspective state and her occurrent perceptual belief about the spin of the electron. This is explicitly spelled out in the state vector $|\Psi'\rangle$, where the three properties are given by three different eigenstates $|\text{"Yes"}\rangle_{h-B}$, $|\text{introspect } e \text{ w}\rangle_{h-PF}$, and $|\text{believes } e \text{ w}\rangle_{h-VC}$ of Hilda's. Measurement of one property of Hilda does not mean deception about another property, as these are separate eigenstates with their own associated operators and eigenvalues. So Hilda is not deceived. There is just a common linguistic output which occurs regardless of the QM interpretation.

In addition, it is important to recall that deception is having a false belief: for X to be deceived about G is for X to believe G when G is not the case. We see that on either side of the superpositions involving Hilda, there are only two beliefs: an introspective belief and a perceptual belief. As deception is a type of belief, these are the only candidates. However, none of these beliefs are instances of deception, because Albert has stipulated beforehand that Hilda "is a competent observer of the positions of pointers" (Albert 77), so her beliefs – perceptual and introspective – about pointer position are accurate. This means any deception in Albert's sense must be at the level of speech output; that is, at the level of Hilda saying "Yes" about introspecting a definite belief about the color of the electron. But speech output, as we have shown, is not an introspective state, and indeed, being an output rather than an intentional state, it is not a belief at all about the color of the electron.

15.5 Conclusion

We have seen that the linearity of operators that represent observables in quantum mechanics leads to cases where a single, common eigenvalue can be elicited from a superpositional state. We have also seen that Albert uses this analysis to claim that an observer of a measurement of a superposition, Hilda, would be deceived about her own mental state. But this claim turns out to be problematic when Hilda's brain states, and in particular her belief states, are analyzed carefully. One problem is that deception is a mental state: a belief with a false content. Deception, therefore, must be exemplified by a belief state. But because the only belief states about measurement available to Hilda are stipulated by Albert to be accurate, then these states cannot be deceptions. Further, spoken output of Hilda's that is common to either possible measurement result in the superposition does not result in Hilda's being deceived, because this output is not itself a form of belief, and in particular, is not an introspective belief as required by the analysis. This means that a competent observer's introspection about her perceptual belief of the measurement of a superposition cannot be a deception.

References

Albert, D. (1992). *Quantum mechanics and experience*. Cambridge, MA: Harvard University Press.

Barrett, J. (1999). *The quantum mechanics of minds and worlds*. Oxford: Oxford University Press.

Brentano, F. (1874). *Psychologie vom Empirischen Standpunkt*. Leipzig: Duncker & Humblot.

Fleming, S., et al. (2010). Relating introspective accuracy to individual differences in brain structure. *Science, 329*(5998), 1541–1543.

Dretske, F. (1988). *Explaining behavior*. Cambridge, MA: MIT Press.

Dretske, F. (1995). *Naturalizing the mind*. Cambridge, MA: MIT Press.

Frege, G. (1892). On sense and reference [Über Sinn und Bedeutung]. *Zeitschrift für Philosophie und Philosophische Kritik, 100*, 25–50.

Lee, T. S., et al. (1998). The role of the primary visual cortex in higher level vision. *Vision Research, 38*, 2429–2454.

Moore, G. E. (1903). The refutation of idealism. *Mind, 12*, 433–453.

Pinker, S. (1994). *The language instinct*. New York: HarperCollins.

Pinker, S. (1997). *How the mind works*. New York: W.W. Norton & Co.

Perry, J. (1977). Frege on demonstratives. *Philosophical Review, 86*, 474–497.

Seymour, K. J., et al. (2016). The representation of color across the human visual cortex: Distinguishing chromatic signals contributing to object form versus surface color. *Cerebral Cortex, 26*, 1997–2005.

Skokowski, P. (1999). Information, belief and causal role. In Moss et al. (Eds.), *Logic, language and computation*. Stanford, CA: CSLI Press.

Skokowski, P. (2018). Temperature, color and the brain: An externalist response to the knowledge argument. *Review of Philosophy and Psychology, 9*(2), 287–299.

Tye, M. (1995). *Ten problems of consciousness*. Cambridge, MA: MIT Press.

Zeki, S. (1993). *A vision of the brain*. Oxford: Blackwell Scientific Publications.

Part III
Quanta and Mind Informs Worldviews

Chapter 16
Absolute Present, Zen and Schrödinger's One Mind

Peter D. Bruza and Brentyn J. Ramm

Abstract Erwin Schrödinger holds a prominent place in the history of science primarily due to his crucial role in the development of quantum physics. What is perhaps lesser known are his insights into subject-object duality, consciousness and mind. He documented himself that these were influenced by the Upanishads, a collection of ancient Hindu spiritual texts. Central to his thoughts in this area is that Mind is only One and there is no separation between subject and object. This chapter aims to bridge Schrödinger's view on One Mind with the teachings of Dōgen, a twelfth century Zen master. This bridge is formed by addressing the question of how time relates to One Mind, and subject-object duality. Schrödinger describes the experience of One Mind to be like a timeless now, whereas subject-object duality involves a linear continuum of time. We show how these differing positions are unified in the notion of 'absolute present', which was put forward in the philosophy of Nishida Kitarō (1871–1945). In addition, we argue that it is in this notion of absolute present that the views of Schrödinger, Dōgen and Nishida meet.

16.1 The Principle of Objectivation

Perhaps you can cast your mind back to your first encounter with the scientific method at school. It may have proceeded something like the following: In the midst of a small group of students a candle is placed and you are asked to light it up. Thereafter you are required to record a series of observations, for example, what is the temperature in the orange part of the flame, the blue part of the flame, or of the wax. At different times you are asked to measure the height of the candle, and record any changes in the colour or consistency of the wax. In this, or other ways, we come

P. D. Bruza (✉)
Faculty of Science and Engineering, Queensland University of Technology, Brisbane, QLD, Australia
e-mail: p.bruza@qut.edu.au

B. J. Ramm
Independent Philosopher, Fremantle, WA, Australia

J. Acacio de Barros, C. Montemayor (eds.), *Quanta and Mind*, Synthese Library 414, https://doi.org/10.1007/978-3-030-21908-6_16

to learn that science involves making observations about some phenomenon, and these observations are input into a scientific understanding of that phenomenon.

It is usually not told, however, what the underlying assumptions are. These are taken as given. First, there is the assumption that the phenomenon is an "object". By this it is meant that the candle has clearly defined boundaries which allows us to isolate and distinguish the candle from non-candle phenomena. In addition, that the candle *is* undoubtedly an object can be seen in the instructions given to the students such as "the object under investigation" etc. An object is assumed to have properties, the values of which can be established by "observation", that is making a measurement of one of the properties, e.g., a particular value of temperature, or a particular colour of the flame. It is also assumed that the properties are independent of each other, for example, by measuring the height of the candle, we don't affect the colour of the wax. Another assumption is that the object persists through time, for example, at different instants we can make record the height of the candle. At each such instant of measurement, there is no doubt that it is the same candle that is being observed and measured.

Usually nothing much is said about the observer. We are simply told to observe the candle, but who or what the observer is remains unknown and somewhat irrelevant to the process. We are told that the observations are "independent of the observer", not much more than that. By observing the candle, we come to tacitly understand that we can peek at the object from the shadows so to speak – as our observations do not influence the object. Finally, we assume that even when we are not observing the candle, it still exists as an object external to ourselves as observers.

Erwin Schrödinger, one of the founders of quantum theory described the principle underlying our candle scenario as "objectivation". By this he meant,

> .. the thing that is also frequently called the hypothesis of the real world around us. I maintain that it amounts to a certain simplification which we adopt in order to master the infinitely intricate problem of nature. Without being aware of it and without being rigorously systematic about it, we exclude the Subject-of-cognizance from the domain of nature that we endeavour to understand. We step with our own person back into the part of an onlooker who does not belong to the world, which by this very procedure becomes an objective world.

This fragment is quoted from an essay (Schrödinger 1992b, p. 118) the goal of which was to elucidate two general principles underlying the scientific method, the "principle of objectivation" being one of the two.

This quote is striking in a number of ways as it seems to contradict some of the assumptions listed above in relation to our observation of the candle. Firstly, the term "hypothesis" is explicitly mentioned in relation to the "real world". As students of science, we are not usually led to hypothesise whether the world around us is real, but as stated above, we simply accept that the candle exists in a reality external to ourselves. Secondly, even though we are taught to be systematic and rigorous when conducting our observations, Schrödinger points out that we are not systematic or rigorous about something else, namely what he calls the "Subject-of-cognizance". At this point we won't go into detail about what the Subject-of-cognizance might be, but simply point out that by assuming the role of the observer of the candle, something of seeming significance is being excluded, and that exclusion is done *without awareness* that something is being excluded. Finally, he states something very extraordinary: By our stepping back "as an onlooker", the world becomes "objective".

16.2 Enter the Subject-of-Cognizance

It would be mistaken to think that Schrödinger's essay was a criticism of the scientific method. Rather, he was trying to clarify the scientific method in the context of Western science and psychology in the early 1940s. Schrödinger states the Subject-of-cognizance as being excluded in the principle of objectivation. By bringing it back in, the philosophical problem of the subject-object relation trails in its wake. There is a relevant historical context in which to consider this age-old problem. While quantum theory was being developed, it seemed that new insights into the subject-object relation may be achieved. The reason for this is that in experiments made on quantum particles, choices made by the observer had a pronounced effect on the quantum-object being observed. It seemed no longer case that the observer (the subject) could peek at the object while residing in the background. For example, Jammer (1974, p. 161) states that Pascal Jordan emphatically stated the very act of observation was *creating* the object being observed. In addition, Jordan seems to have at least considered subjectivism, "we ourselves produce the results of measurement" [Wir selber rufen die Tatbestände hervor] (Jordan 1934, p. 228). Despite some of the founders of quantum theory leaning toward subjectivism, it made many physicists, both then and now, decidedly uncomfortable. In the meantime it has been well documented how the field of quantum physics circumvented subjectivism by means of the Copenhagen interpretation (e.g., see Rosenblum and Kuttner 2006).[1]

Schrödinger's position in regard to the subject-object relation was not a circumvention, but radically to the point. At the very end of his essay he states (Schrödinger 1992b, p. 127):

> All this was said from the point of view that we accept the time-hallowed distinction between subject and object. Though we have to accept it in everyday life "for practical reference", we ought, so I believe to abandon it in philosophical thought... *The world is given to me only once, not one existing and one perceived. Subject and object are only*

[1]Chalmers and McQueen (2014) have recently argued in support of consciousness playing a causal role in collapsing the wave function. They posit the existence of m-properties which can never be superposed. Whenever a superposed property becomes (potentially) entangled with an m-property it necessarily collapses into a definite state. Consciousness is a natural candidate for an m-property since it cannot be superposed; for example, I cannot experience red and not-red simultaneously in the same location. This hypothesis has the theoretical virtues of: (1) providing a potential solution to the measurement problem, in particular, an explanation for why measurement collapses the wave function into a definite state, and (2) gives consciousness a causal role in the physical world. Schrödinger goes further than Chalmers and McQueen's dualist theory by holding that the physical world is in fact constituted by conscious experience.

one. The barrier between them cannot be said to have broken down as a result of recent experience in the physical sciences, for this barrier does not exist [emphasis added]

Schrödinger's mention of "recent experience in the physical sciences" alludes to the previously mentioned reconsideration of the problem of the subject-object relation raised by quantum physics. Schrödinger's stance is very unorthodox in two respects: (1) there is *no duality* between subject and object (the orthodox position is that there is an inherent distinction) and (2) the only world that is "given" is the one that is actually being perceived. This runs counter to the orthodox view that the world exists independent of our perception of it.

16.3 Zen and One Mind Subjectivism

> A perception sudden as blinking, that subject and object are one, will lead to a deeply mysterious wordless understanding; and by this understanding you will awake to the truth of Zen

This is a quote from Huang Po (d 850) a well-known master from the Chinese period of Zen known as Chan (Blofeld 1958). The essence of Huang Po's quote is the same as Schrödinger's, namely that subject and object are one in a spontaneous act of perception. Moreover, Chan teaches that this perception is a discontinuity in the sense that it is not simply revealing a pre-existing non-dualistic form of reality. More specifically similar to Huang Po's "perception sudden as blinking", Schrödinger wrote in a letter to Bruno Bertotti, one of his students (Bertotti 1985):

> ...the very, very old Indian TAT TWAM ASI (This art thou) [subject-object are one] is, of course, not a physical but rather a metaphysical statement. It is so simple that it is impossible to explain it. It cannot be grasped by the intellect, *but it may spring up in you at some occasion like a spark*, and then it is there and will never really leave you, even though it is not a practical maxim to use every hour of your life. [emphasis added]

In the history of Zen there are many stories of such spontaneous experiences. They are often expressed in relation to perception via a particular sense modality, for example, the sense of sound.

> Monk: Where can I enter Zen?
> Master Gensha: Can you hear the babbling brook?
> Monk: Yes, I can hear it.
> Master Gensha: Then enter it there.

The oneness of subject and object is a holistic experience of there being no distinction between that which is hearing (subject), the act of hearing, and that which is heard (the object). Such experiences directly undermine the distinction and assumed separation between an observer (subject) perceiving a candle (object).

If subject and object are one, what then is doing the observing? This is a very important question in Zen. It is common to assume the subject, the one doing the observing, is a product of consciousness in an individual's brain, and it would therefore be understandable to define Schrödinger's Subject-of-cognizance in this way. However, he himself did not adhere to such a definition. Schrödinger rejected this theory for two reasons. Firstly, because "the stuff from which our world picture is built, is yielded exclusively from the sense organs as organs of the mind, so that every man's world picture is and always remains a construct of his mind and cannot be proved to have any other existence." (Schrödinger 1992a, p. 122). Unsurprisingly then, the mind itself cannot be found within this construct. He points out secondly, that no one has ever observed consciousness in the head or body. The scientist rather finds only "millions of cells of very specialized build in an arrangement that is unsurveyably intricate"(ibid., p. 123). He goes on "nowhere you may be sure, however, far physiology advances, will you ever meet the personality, will you ever meet the dire pain, the bewildered worry within this soul." (ibid., p. 124). In summary, we cannot find mind in the world because our world picture is constructed by the mind. Moreover, "The reason why our sentient, percipient and thinking ego is met nowhere within our scientific world picture can easily be indicated in seven words: because it is itself that world picture. It is identical with the whole and therefore cannot be contained in it as a part of it." (Schrödinger 1992a, p. 128). The mind is thus the container of the world picture, not something within that picture (ibid., p. 136–137).

We have seen examples of Scrödinger's philosophical method, which is in the spirit of 'radical empiricism' as employed by William James (1976). Based upon experience he rejected the separability of subject and object, and hence the hypothesis of an independent world. This method, combined with philosophical arguments further lead him accept the conclusion that rather than many consciousnesses, there was in fact only One Mind, and hence one observer. In fact, Schrödinger stated in the 1950s that his world view was formed by Spinoza and Schopenhauer. In particular, the view that he formed when he was 30-years old, and never changed, was that of the One Mind presented in the Upanishads (Bitbol 1999).

Bruno Bertotti was one of Schrödinger's students and was clearly intimate with his views (Bertotti 1985). Bertotti's defined of Schrödinger's world view as "rational mysticism" and summarized it as follows: "his main philosophical theme, namely that all existents, in particular the individual consciousnesses, are manifestations of a Single Mind".

Why did he hold this view? In another essay, Schrödinger refers to inevitable paradoxes that spring from a single source which he termed the "arithmetic paradox" (Schrödinger 1992a). He defines this paradox as "The many conscious egos from whose mental experiences the one world is concocted". More specifically (Bertotti 1985),

> The most wide-spread attitude is, so I believe, the following: there is one real world and this naturally accounts for its making the same impression on Mr. A, Mr. B, etc., etc .. to me it seems the greatest and absolutely inexplicable marvel that "we all live in the same world"... Why marvel? Well, you see, my world is built up of my sensations, the world of Mr B is built up of Mr B's sensations. There is absolutely no communication between Mr B's sensations and mine... Is it not then actually an unaccountable marvel that these "two worlds", built as it were from entirely different material, coincide?

It is apparent from the preceding that Schrödinger did not subscribe to the common sense view that the correspondence between the sensations of Mr. A and Mr. B. is due to them both perceiving the same external reality. He rejected this doubling of the world into appearance and reality, and any empirically suspect notions such as Kant's thing-in-itself (Schrödinger 1992a, p. 127), hence this solution was unacceptable to him. He also rejected Leibniz's "fearful doctrine of monads" (ibid., p. 129). Although Leibniz's view avoided mysterious things in themselves and did not make the error of locating consciousness in things, it entailed a splittering of the universe into numerous non-interacting worlds.

Schrödinger's resolution of the arithmetic paradox once again involved adopting a radically unorthodox position (ibid):

> There is obviously only one alternative [to the paradox], namely the unification of minds, or consciousnesses. Their multiplicity is only apparent, in truth there is only one mind. This is the doctrine of the Upanishads. And not only of the Upanishads. The mystically experienced union with God regularly entails this attitude.

Schrödinger's view is there is only "one mind", not individual consciousnesses. Therefore, the only possible observer is this One Mind. Finally, Schrödinger, again employing a method of radical empiricism argued:

> The doctrine of identity (of all minds) can claim that it is clinched by the empirical fact that consciousness is never experienced in the plural, only in the singular. Not only has none of us ever experienced more than one consciousness, but there is also no trace of circumstantial evidence of this ever happening anywhere in the world. (ibid, p. 130)

This observation does not logically entail his conclusion since an absence of evidence of multiple consciousnesses is not the same as evidence of absence of multiple consciousnesses. Nevertheless, as an empirical argument, no evidence for something (for example unicorns, Homerian gods etc.) provides a reason for not believing in that thing. Furthermore, the phenomenology was not meant to stand alone, but to work in conjunction with his other philosophical and empirical arguments.

Did Schrödinger ever attempt to incorporate his world view based on the One Mind into his scientific theories? Bertotti was unable to uncover any tangible attempt, however in Bertotti's description of Schrödinger's vision, he does allude to the possibility of there being one (Bertotti 1985):

> On the one hand the essential unity of the world arises from its fragmentation among the individual consciousnesses not because of a cogent philosophical argument, but through a 'mystical' awareness; on the other hand this very 'oneness', being independent from the single observers, makes quantum mechanics unacceptable and *urgently requires a new theory*. [emphasis added]

The question left dangling is whether a scientific theory could be developed around the One Mind. Such a question would seemingly need to bridge Eastern philosophy and the realism of contemporary science. The rest of this chapter will not attempt to present such a bridge, but rather focus on the concept of time – a concept which is core to both sides of the bridge.

16.4 Subject, Object, Time, NOW

Let us go back to the scenario of observing the candle. Recall that one of the tacit assumptions was that the object (the candle) persists through time and the observer can make observations at various time instants. Regardless of having "stepped back as an onlooker", it is reasonable to assume that the observing subject similarly persists through time as illustrated by Fig. 16.1. This diagram represents two states of the observer, where, for example, +1 represents the observer is in the state attending to the candle and −1 represents a state in which the observer has become distracted. In this way the state of the observing subject can be modelled as having a linear temporal trajectory through a space of two possible states. As mentioned in the introduction, a feature of this time-based mode of observation is a seeming separation between observer (subject) and observed (object), which both Huang Po and Schrödinger deny. Contrast this time-based account with Huang Po's "a perception sudden as blinking that subject and object are one", which suggests the experience of One Mind is a discontinuity. This notion of discontinuity also accords with Schrödinger's observation mentioned earlier "…it [one mind awareness] may spring up in you at some occasion like a spark". But does such a state of One Mind persist through time?

In his philosophical exploration of time Schrödinger (1992c) inquires into whether science and philosophy can illuminate religious beliefs about life and death and the transcendence of time. He approvingly interprets Kant as having proposed that:

> To be spread out in space and to happen in a well-defined temporal order of 'before and after' is not a quality of the world that we perceive, but pertains to the perceiving mind which, in its present situation anyhow, cannot help registering anything that is offered to it according to these two card-indexes, space and time. (p. 144)

Schrödinger again poses doubt as to whether space and time are qualities of an objective world. He asks, even if we do accept the hypothesis of an objective world, on what basis are we to decide whether an experienced feature belongs to our mind or the world (ibid, p. 145–146)? This being said, the most insightful point to take away from Kant, according to Schrödinger, comes from Schopenhauer's reading of Kant:

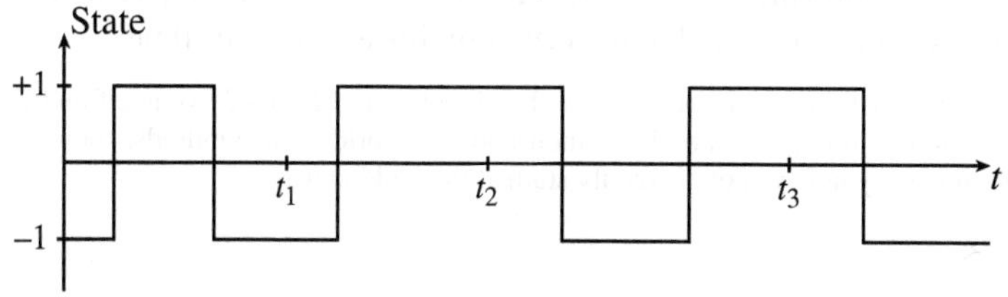

Fig. 16.1 The state of awareness of an observing subject

> The great thing was to form the idea that this one thing 'mind or world' may well be capable of other forms of appearance that we cannot grasp and that do not imply the notions of space and time. This means an imposing liberation from our inveterate prejudice. There are probably other orders of appearance than the space-time like. (ibid., p.146)

What would an experience beyond space and time even be like, and how does it apply specially to One Mind awareness? Once again Schrödinger answers this question from direct experience:

> Mind is by its very nature a *singular tantum*. I should say: the over-all number of minds is just one. I venture to call it indestructible since it has a peculiar time-table, namely mind is always *now*. There is really no before and after for mind. There is only a now that includes memories and expectations (Schrödinger 1992a, p. 135).

The state of One Mind, then, according to Schrödinger, is experienced as a kind of timeless NOW. He denies that 'before' and 'after' apply to the experience of the present moment. Rather 'before' and 'after' are constructed in the timeless NOW by memory and expectation.

Schrödinger's phenomenological descriptions of this timeless present moment are echoed in spiritual traditions, with a particular affinity with Zen Buddhism, as is evident from D. T. Suzuki's summary of Zen:

> Zen is emphatically a matter of personal experience; if anything can be called radically empirical it is Zen. No amount of reading, no amount of teaching, no amount of contemplation will ever make one a Zen master. Life itself must be grasped in the midst of its flow; to stop it for examination and analysis is to kill it, leaving its cold corpse to be embraced. (Suzuki 1954, p. 132)

Dōgen, a twelfth century Zen master, provides descriptions of temporal experience that are particularly relevant to those of Schrödinger. Dōgen[2] (1200–1253) was one of the most prolific writers amongst Zen masters of old and contributed an extensive collection of articles on various topics (Tanahashi 2010a,b). The chapter of relevance to the present discussion is titled "The time being" (sometimes translated as "Being time" (Dōgen 2010).

> The time being has a characteristic of flowing. So-called today flows into tomorrow, today flows into yesterday, yesterday flows into today. And today flows into today, tomorrow flows into tomorrow. Because flowing is a characteristic of time, *moments of past and present do not overlap or line up side by side* [emphasis added]

This position aligns with Schrödinger's because it denies a linear succession of time instants. In the following passage, Dōgen also alludes to the subject-object mode of awareness as being connected to the orthodox linear notion of time:

> In your study of flowing, if you imagine the objective to be outside yourself and that you flow and move through hundreds of thousands of worlds, for hundreds, thousands and myriad of eons, you have not devotedly studied the buddha way.

[2]Zen master Dōgen was the founder of the Soto sect, one of the two main schools of Zen Buddhism in Japan.

Here Dōgen implies that it is illusory to assume an objective universe through which an individual moves through space and time. The striking element here is that the objective universe seems so ordinarily real, so real that it doesn't bear questioning. In Zen, however, the objective universe is viewed as an almost perfect illusion where "devotedly studying the buddha way" involves earnestly questioning the assumed reality of the objective universe to directly understanding the reality of this illusion. This understanding is a not a conventional understanding, but rather, as Huang Po states "a wordless understanding", which is brought about by experiencing One Mind awareness. Such a radically subjective understanding of space and time as found in Zen is again reflected in Schrödinger's writings when he says about the conscious mind that:

> It is the stage, and the only stage on which this whole world process takes place, or the vessel or container that contains it all and outside which there is nothing. (Schrödinger 1992b, p. 136).

There is one philosopher who is uniquely positioned to bridge Dōgen's view of time and Schrödinger's view of One Mind. His name is Nishida Kitarō. Nishida trained in Zen Buddhism and according to D.T. Suzuki, an authoritative Zen scholar, Nishida had an enlightenment experience which deeply influenced his views about the ultimate nature of reality. In the context of this article, Nishida's experience can be construed as a direct experience of Shrödinger's One Mind. Rather than write from a spiritual perspective like the ancient Zen masters, Nishida was determined to write about the essence of reality in a way that would be acceptable to the contemporary philosophy. With this choice, Nishida was unknowingly subscribing to Schrödinger's rational mysticism in the sense that his philosophy presents itself as a rationalism that proposes to analyze reality which has been grasped on the basis of a profound religious, or spiritual experience.

Whilst Nishida does not use the term One Mind, he does directly refer to it in terms of the self as a borderless "absolute present". He states (Raud 2004),

> When I speak of ourselves being singular focal points of the world determining our individualities through self expression, this does not mean that I conceive of the self necessarily in terms of the logic of objects. It is rather, a singular centre of the absolute present that includes in itself the eternal past and future. This is why I call the self a momentary self determination of the absolute present ... And the world of the absolute present is the sphere with infinite radius and no circumference, which has a center everywhere.

Nishida's view refers to the 'logic of objects'. By this he meant something very similar to the Schrödinger's principle of objectivation, namely that the logic of objects occurs when the subject steps back as an observer thus rendering a universe of objects. Moreover, Nishida's 'absolute present' resonates with Schrödinger's notion of the timeless NOW, but in Nishida's view, this 'absolute present'[3] also *includes* both future and past. This view clearly aligns with Dōgen's, "today flows into tomorrow, today flows into yesterday, yesterday flows into today...".

[3]Nishida also uses the expression 'eternal now' elsewhere in his writings.

In contrast to the 'logic of objects', Nishida proposed the 'logic of absolutely contradictory self-identity'. This logic is the foundation of Nishida's philosophy, which was deeply influenced by his practice of Zen (Kozyra 2018). Kozyra (2018, p. 432) describes Nishida's logic as when "the subject transcends itself and perceives the object 'through becoming the object'. We perceive the world in our absolutely contradictory self-identity with the world". This definition clearly has the same essence as Schrödinger's One Mind in which subject and object are one.

Nishida's term 'absolute present' may seem like a term which is time-based, but Nishida's quote above also importantly refers to space, which he describes as boundless and without a center. Raud (2004, p 44.) interprets the "center" as equating to an individual subject or 'conscious self'. He further argues that the dichotomy of space and time obtains from the standpoint of such a self. This dichotomy is transcended when these time and space unite in 'contradictorily self-indexical' point of view. Raud stresses that the unification should not be posited as being outside the context of reality, because when the self determines itself to be the 'absolute present', then the self is none other than reality itself. This reality paradoxically manifests a borderless space in which seemingly distinct phenomena have no separation and is a timeless NOW which includes time. It is at the 'absolute present', Schrödinger, Dōgen and Nishida converge.

16.5 Closing Reflections: The Illusion of Time's Arrow

Here we have investigated Schrödinger's views on subject-object non-duality and the One Mind hypothesis. We also showed how he questioned subject-object duality by denying that an observation is the relation of the objective time of an object and the separate subjective states of an observer. For Schrödinger space and time, if they are anywhere at all, belong to the observer-observed unity (not outside of it). Schrödinger also sought for a liberation from time through philosophy and mystical modes of experience, particularly in the timeless NOW. As a scientist, however, he was not content with these philosophical and mystical approaches by themselves. His thinking was also constrained by the findings of the sciences. We would like to close by asking: how do his reflections on time accord with physics?

In fact, Schrödinger believed that the non-objectivity of time is also supported by physics. In his thoughts on science and religion (Schrödinger 1992c), he grappled with the enduring mystery of the arrow of time. The problem is that for observers, time seems to flow uniformly in a single direction from past to future. Yet the laws of mechanics are symmetrical (Schrödinger 1992c, p. 151). The arrow of time is not fundamental to these laws. As a consequence, there is no reason to believe that the world is not like a film that could be played backwards as well as forwards. To explain this, Schrödinger draws upon Boltzmann's statistical theory of time. Boltzmann demonstrated that the arrow of time could be explained by the second law of thermodynamics that a system tends towards greater entropy. That is, time is not physically fundamental in this view, rather the reason that eggs do not unbreak is not because they cannot, but because it is simply a very unlikely event. For Schrödinger the statistical theory of time's arrow provided further support for Kant's theory that time is subjective. There is however a tension that Schrödinger does not resolve, namely in the formalization of quantum theory, time has both an objective and fundamental treatment.

Interestingly, however, a number of recent studies have experimentally linked the thermodynamic arrow of time to quantum theory. One study showed that a sub-system tends towards equilibrium because of quantum entanglement of its particles with surrounding systems (Linden et al. 2009). This means that a cup of coffee will tend to cool down (reach equilibrium) because of the entanglement of its particles with surrounding air particles. Another study demonstrated that the thermodynamic arrow of time could be reversed when two particles were initially quantum mechanically correlated (Micadei et al. 2017). This shows that the thermodynamic arrow of time is relative to initial conditions, it is not an absolute. The important point of these findings is that time is not a background flow that things happen in, but is reducible to the mechanics of the systems themselves. It is an open question whether time will eventually be eliminated from all fundamental physical theories, including quantum theory. Important steps in this direction have been made by Julian Barbour.

Barbour (1999) assumes that time is a derivable notion based on change. He presents a theory based on a configuration space $\mathcal{U}$, each point of which is a particular configuration of all matter in the universe (called a 'now'). A path between two points is a curve where time is derived from differences between the configurations corresponding to the points on the curve. One of the consequences of Barbour's theory is that it parsimoniously supports a timeless Newtonian dynamics (Barbour 2009).

When Schrödinger penned his views about One Mind, his exalted place in the history of science had already been assured. Nevertheless, his unorthodox views on such matters were politely ignored, and even today these views are relatively unknown. In the mean-time quantum theory has become the most stunningly successful theory ever devised by humans. To date, quandaries raised by quantum theory such as subject-object duality remain unresolved. There are some small signs, however, that such foundational issues are being revisited. For example, Malin (2003) has written an intriguing account of the foundations of quantum theory from a Western philosophical perspective. What is very striking about this book is its third part titled "Physics and the One". Following from Schrödinger, Malin argues that the next step for science is to transcend subject-object duality. In this regard, Malin issues a thought provoking and daring question: "The quest for the One Mind calls for the transcendence of the subject/object mode. Can science participate in this quest?".

Acknowledgements The first author respectfully dedicates this article to Professor Shimon Malin, who made the first author aware of Schrödinger's views on One Mind as well as for his thought provoking views about quantum theory. The first author also thanks Michel Bitbol for the inspiring discussions in the subject area of this article.

References

Barbour, J. (1999). *The end of time: The next revolution in physics*. Oxford/New York: Oxford University Press.

Barbour, J. (2009). *The nature of time*. arXiv:9003.3489v1.

Bertotti, B. (1985). The later work of E. Schrödinger. *Studies in History and Philosophy of Science, 16*(2), 83–100.

Bitbol, M. (1999, August). *Schrödinger and Indian philosophy*. http://michel.bitbol.pagesperso-orange.fr/Schrodinger_India.pdf. (Online; Accessed September 06, 2018)

Blofeld, J. (1958). *The Zen teaching of Huang Po*. New York: Grove Press.

Chalmers, D., & McQueen, K. (2014, May). *Consciousness and the collapse of the wave-function.* A lecture given by Davis Chalmers at Göttingen University. From: http://ieet.org/index.php/IEET/more/chalmers20140806

Dōgen. (2010). The time being. In K. Tanahashi (Ed.), *Treasury of the true Dharma eye: Zen master Dōgen's Shobo Genzo* (Vol. 1, pp. 104–111). Boston: Shambhala.

James, W. (1976). *Essays in radical empiricism.* Cambridge: Harvard University Press.

Jammer, M. (1974). *The philosophy of quantum mechanics: The interpretations of quantum mechanics in historical perspective*. New York: Wiley.

Jordan, P. (1934). Quantenphysikalische bemerkungen zur biologie und psychologie. *Erkentniss, 4*, 215–252.

Kozyra, A. (2018). Nishida Kitarōs philosophy of absolute nothingness (zettaimu no tetsugaku) and modern theoretical physics. *Philosophy East and West, 68*(2), 423–446.

Linden, N., Pospescu, A., Short, A., & Winter, A. (2009). Quantum mechanical evolution towards thermal equilibrium. *Physical Review E, 79*(6), 061103.

Malin, S. (2003). *Nature loves to hide: Quantum physics and reality, a Western perspective*. New York: Oxford University Press.

Micadei, K., Peterson, A., Souza, R., Sarthour, I., Oliveira, G., Landi, T., & Lutz, E. (2017). *Reversing the thermodynamic arrow of time using quantum correlations.* arXiv:1711.03323.

Raud, R. (2004). 'Place' and 'Being-Time': Spatiotemporal concepts in the thought of Nishida Kitarō and Dōgen kigen. *Philosophy East & West, 54*(1), 29–51.

Rosenblum, B., & Kuttner, F. (2006). *Quantum enigma: Physics encounters consciousness.* Oxford/New York: Oxford University Press.

Schrödinger, E. (1992a). The arithmetic paradox: The oneness of mind. In E. Schrödinger (Ed.), *What is life? With mind and matter and autobiographical sketches* (pp. 128–139). Cambridge/New York: Cambridge University Press.

Schrödinger, E. (1992b). The principle of objectivation. In E. Schrödinger (Ed.), *What is life? With mind and matter and autobiographical sketches* (pp. 117–127). Cambridge/New York: Cambridge University Press.

Schrödinger, E. (1992c). Science and religion. In E. Schrödinger (Ed.), *What is life? With mind and matter and autobiographical sketches* (pp. 128–139). Cambridge/New York: Cambridge University Press.

Suzuki, D. (1954). *An introduction to Zen buddhism.* New York: Grove Press.

Tanahashi, K. (Ed.). (2010a). *Treasury of the true Dharma eye: Zen master Dōgen's Shobo Genzo* (Vol. 1). Boston: Shambhala.

Tanahashi, K. (Ed.). (2010b). *Treasury of the true Dharma Eye: Zen master Dōgen's Shobo Genzo* (Vol. 2). Boston: Shambhala.

Chapter 17
Semantic Gaps and Protosemantics

Benj Hellie

Abstract Semantic gaps between physical and mental discourse include the 'explanatory', 'epistemic' (Black-and-White Mary), and 'suppositional' (zombies) gaps; protosemantics is concerned with what is fundamental to meaning. Our tradition presupposes a truth-based protosemantics, with disastrous consequences for interpreting the semantic gaps: nonphysicalism, epiphenomenalism, separatism. Fortunately, an endorsement-based protosemantics, recentering meaning from the world to the mind, is technically viable, intuitively more plausible, and empirically more adequate. But, of present significance, it makes room for interpreting mental discourse as expressing simulations: this blocks the disastrous consequences; and, as a bonus, accommodates hitherto anomalous asymmetries among the various semantic gaps.

A 'semantic gap' divides two regions of language when the meanings on the one side are very different from those the other. The gap considered here divides language about the physical (quotidian or scientific, classical or quantum alike) from language about the mental—as discussed in an extensive philosophical literature over the last half-century (perhaps continuously with much older literature), which attempts to establish its existence and interpret its significance. To my mind, the gap (if perhaps not each of its alleged manifestations) genuinely exists, but the literature misinterprets its significance.

The source is a pervasive, but fundamentally mistaken assumption regarding the 'protosemantical' issue of what meaning *is* (essentially or fundamentally): roughly, meaning is assumed to be *representation* (depiction, description, information); sharpening, a theory of meaning is assumed to be ineliminably a theory of *truth-conditions* imposed on *the world*. While pervasive and venerable, the assumption is at best viable in a local special case, at worst a gross oversimplification resulting from a misunderstanding of that special case—or, at any rate, it is surely

B. Hellie (✉)
Department of Philosophy, University of Toronto, Toronto, ON, Canada
e-mail: benj.hellie@utoronto.ca

J. Acacio de Barros, C. Montemayor (eds.), *Quanta and Mind*, Synthese Library 414, https://doi.org/10.1007/978-3-030-21908-6_17

optional. For on a viable alternative, meaning is, roughly, *expression* (display, commitment, articulation); sharpening, a theory of meaning is ineliminably a theory of *endorsement-conditions* imposed on *our mental states*.

Endorsement-conditionalism relocates the 'perspective' of meaning from the world to the mind, marking a radical departure from our tradition's long-hegemonic truth-conditionalism (but also from a far older broadly 'scientistic' outlook, over-emphasizing 'objectivity' and downplaying 'empathy'). Theory of meaning benefits extensively; even better, the startup cost is minimized by extensive continuity with central truth-conditionalist apparatus. Significant for present purposes, though, is the very different take it offers on semantic gap phenomena: to my mind, a strongly preferable one, free from the spectres of dualism, epiphenomenalism, and a disintegrating 'separatist' image of mind.

Section 17.1 sketches the various semantic gap phenomena—the 'explanatory gap'; the 'epistemic gap' (exemplified by 'Black-and-White Mary'); the 'suppositional gap' (exemplified by the alleged 'conceivability' of zombies). Section 17.2 gives the sense in which these are 'semantic', by connecting them to *logical consequence* and thence *meaning*; sketches up a more-or-less state of the art framework for truth-conditionalist semantics; and with it extracts from the semantic gap phenomena various *dispiriting consequences* (antiphysicalism; epiphenomenalism; 'separatism'). Section 17.3 sketches the endorsement-conditionalist alternative, and implements an 'expressivist' analysis of mental sentences in close alliance with 'simulationism' (on which reasoning about the mental involves 'empathy', analyzed here as akin to *supposition*). Section 17.4 offers the endorsement-conditionalist take on the semantic gaps: first, the 'expressivism' blocks the *dispiriting consequences*; second, various details of the account are used to analyze the various semantic gap phenomena—in particular, epistemic gaps are traced to imperfections of empathy; the 'trivalence' behind endorsement-conditionalism severs the path from epistemic to suppositional gaps, and both zombies and inverts are shown to be inconceivable; and the explanatory gap is traced to an essential 'viewpoint-shift' between reasoning about the physical and about the mental.

17.1 Semantic Gap Phenomena

Various (purported—to avoid archness, the qualification is left tacit through this section) semantic gap phenomena have been canvased in recent literature (some, perhaps, continuous with older discussion): adapting the overview/systematization in Chalmers (2002a, 3.1–3.3),[1] I sketch three distinctive exemplars.

[1]Terminology diverges somewhat: Chalmers discusses an 'explanatory argument', a 'conceivability argument', and a 'knowledge argument', grouping my semantic gaps as 'epistemic gaps'—somewhat inappositely, as 'conceivability' (and to some extent, 'explanation') is obscurely related to the 'epistemic', while 'knowledge' seems to exhaust it.

A *first* semantic gap phenomenon is the *explanatory gap*, involving mental facts left unexplained (in a 'constitutive' sense) by the totality of physical fact. An early statement is Leibniz's famous 'mill' example:

> [P]erception and that which depends upon it are inexplicable on mechanical grounds, that is to say, by means of figures and motions. And supposing there were a machine, so constructed as to think, feel, and have perception, it might be conceived as increased in size, while keeping the same proportions, so that one might go into it as into a mill. That being so, we should, on examining its interior, find only parts which work one upon another, and never anything by which to explain a perception. (Leibniz 1714/1991, 17)

Contemporary discussion of the explanatory gap is initiated (and its name bestowed) by Levine (1983), who observes that 'what is left unexplained by the discovery of C-fiber firing is *why pain should feel the way it does*! For there seems to be nothing about C-fiber firing which makes it naturally 'fit' the phenomenal properties of pain, any more than it would fit some other set of phenomenal properties' (357); the explanatory gap is central to Chalmers's early paper ('why should physical processing give rise to a rich inner life at all? It seems objectively unreasonable that it should': Chalmers 1995, 5) announcing the 'hard problem' of resolving the follow-on explanatory challenge. Chalmers's overview puts it this way: 'even once one has an explanation of all the relevant functions in the vicinity of consciousness—discrimination, integration, access, report, control—there may still remain a further question: why is the performance of these functions accompanied by experience?' (Chalmers 2002a, 248).

Second is the *epistemic gap*, involving mental facts for knowledge of which knowledge of the totality of physical fact is insufficient. A relatively early statement is Broad's discussion of a 'mathematical archangel':

> He [the archangel] would know exactly what the microscopic structure of ammonia must be; but he would be totally unable to predict that a substance with this structure must smell as ammonia does when it gets into the human nose. The utmost that he could predict on this subject would be that certain changes would take place in the mucous membrane, the olfactory nerves and so on. But he could not possibly know that these changes would be accompanied by the appearance of a smell in general or of the peculiar smell of ammonia in particular, unless someone told him so or he had smelled it for himself. (Broad 1925, 71)

Contemporary literature contains two particularly high-profile discussions: Nagel (1974) bemoans that, although 'I want to know what it is like for a *bat* to be a bat' (439), barriers well beyond mere physical ignorance impede his satisfaction; Jackson (1982) introduces 'Black-and-White Mary', the color scientist trapped for life in a monochrome environment: released to see color, 'it is just obvious that she will learn something[]. But then [] her previous knowledge was incomplete. But she had *all* the physical information' (130). Chalmers's overview affirms Jackson's assessment: 'Despite all her knowledge, [] Mary [] does not know what it is like to see red. Even complete physical knowledge and unrestricted powers of deduction do not enable her to know this' (Chalmers 2002a, 250).

Third is the *suppositional gap*, involving mental facts coherently supposed absent despite preserving the totality of physical fact. An early statement is Descartes's claim that 'while I could pretend that I had no body and that there was no world and no place for me to be in, I could not for all that pretend that I did not exist' (Descartes 1637/1985, 127). Contemporary literature favors the opposite direction of independence: 'Descartes's argument also has the following turned-around version, which to my knowledge he never employed. The existence of the body without the mind is just as conceivable as the existence of the mind without the body' (Nagel 1970, 401); compare also Kripke (1980, 148–54). Chalmers's overview relays such reports: where a 'zombie' is 'a system that is physically identical to a conscious being but that lacks consciousness entirely', 'many hold that [] we can coherently imagine zombies, and there is no contradiction in the idea that reveals itself even on reflection' (Chalmers 2002a, 249; a similar point is made for 'inverts', possessing consciousness, but of a very different sort than corresponding actual subjects).

17.2 Semantic Gaps as Truth-Conditional Gaps

I call these *semantic* gaps, because each of the involved phenomena—explanatory completeness; the transmission of knowledge; constraints on supposition—is closely related to phenomena of meaning. Starting from the *meaning* side of the relation, note that, very plausibly, the meanings of a set of 'premiss' sentences and a 'conclusion' sentences entirely determine whether the premisses *entail* the conclusion (have it as a 'logical consequence'). Next, consider the notion of *endorsing* a sentence, understood as carrying an 'implicit commitment' to 'accept' the sentence (because of the meaning the sentence has for one and the specifics of one's mental state). Note last that each of these equivalences is quite plausible:

I. φ entails ψ just if endorsing φ requires endorsing ψ (on pain of unintelligibility)
II. Endorsing φ requires endorsing ψ just if:

   A. Granting φ forecloses *wondering why* (seeking any further explanation for) ψ
   B. *Knowing* φ (endorsing φ 'with knowledge') requires knowing ψ
   C. *Supposing* φ (endorsing φ within a suppositional state) requires supposing ψ

An explanatory, epistemic, or suppositional gap, by (II), exists just in the presence of an 'endorsement gap'; by (I), this exists just in the absence of entailment; and the presence or absence of entailment, as noted, is entirely determined by (certain facts about) meaning: so the various gaps follow in lockstep with those facts about meaning (whatever they may be)—making them manifestations of an underlying *semantic* gap.

A systematic account and evaluation of the various gaps therefore requires understanding those underlying facts about meaning—calling quite naturally for a turn to theory of meaning. Philosophical literature has for a half-century accorded paradigm status to *truth-conditional* theories of meaning, making it helpful to review certain important but intricate details of the internal structure of such theories. The founding idea is the Frege-Tarski analysis of logical consequence as the preservation of truth under all conditions—a set of premisses Ψ entails a conclusion φ just if for any condition, if every member of Ψ is true in that condition, then φ is also true in that condition. Because, moreover, facts about entailment flow exhaustively from facts about meaning, the meaning of a sentence must determine the conditions under which it is true; and hence a theory of meaning for a language must state the truth-conditions of its sentences.

A long shadow is still cast by Davidson's early implementation (Davidson 1967): it has two key components. First, sentence-meanings are portrayed with *T-sentences*, disquotational biconditionals stating truth-conditions: *'goats eat cans' is true just if goats eat cans*; or, more flexibly, *'it rains' is true relative to condition k just if it rains in condition k*. Second, a core explanatory ambition is *compositionality*, with the meaning of a complex expression determined by the meanings of its constituents—so by the first component, whenever φ and ψ share a truth-condition (their T-sentences share a right hand side), then so do any sentences $\Phi(\varphi)$ and $\Phi(\psi)$ differing only in intersubstitution of φ and ψ.

This early implementation is inadequate. Observe that it rules out pairs of logically equivalent sentences φ and ψ with some inequivalent $\Phi(\varphi)$ and $\Phi(\psi)$: by the Frege-Tarski basis, logical equivalence is identity of truth-condition, so equivalent φ and ψ share a truth-condition; so, by compositionality, so do any $\Phi(\varphi)$ and $\Phi(\psi)$; so by the Frege-Tarski basis, $\Phi(\varphi)$ and $\Phi(\psi)$ are likewise equivalent. Unfortunately, such pairs appear to exist (or at least theory should make room for them)—canonically, *it rains* and *now, it rains* are equivalent; *always, it rains* and *always, now, it rains*, inequivalent (locus classicus: Kamp 1971).

One repair distinguishes *compositional* from *logical* truth-conditions (if the latter abstract from the former, φ and ψ can share logical but not compositional truth-conditions, lifting the requirement that $\Phi(\varphi)$ and $\Phi(\psi)$ share compositional truth-conditions or therefore logical truth-conditions). Unfortunately, yet a third notion of 'truth-condition' is required upon expanding the 'client base' for theory of meaning beyond logic, to include *pragmatics* (concerned with the use of sentences in assertions to convey information, or with stative attitudes toward sentences, particularly *endorsement*, to 'package' information for eventual use in assertion or reasoning). After all, compositionally distinct sentences (for example, *it rains* and *now, it rains*) can invariably convey the same information—so the compositional truth-condition is distinct from any 'pragmatical truth-condition'. But such a 'pragmatical truth-condition' is also distinct from the logical truth-condition, for logically equivalent sentences can convey distinct information (canonically, just as *either (always, now, it rains) or (always, it is not the case that now, it rains)* is valid, so too is *either (necessarily, actually, it rains) or (necessarily, it is not the case that actually, it rains)*—namely, *it is* ***noncontingent whether*** *actually, it rains* is valid: and if so, *actually, it rains* should invariably convey noncontingent information; but *it rains* invariably conveys contingent information; and yet *it rains* and *actually, it rains* are logically equivalent).

The mutual relevance of these proliferating truth-conditions is resolved in a *Standard Framework* developed over the course of the 1970s (Lewis 1970, 1980; Stalnaker 1970, 1978; Kaplan 1977):

S1. 'Compositional truth-conditions' for sentences are replaced with a more general notion of the *semantic value* of an expression: an abstract entity assigned to a simple expression by the meaning-bestowing conventions of language and to a complex expression by the composition of semantic values of its constituent expressions;

S2. 'Pragmatical truth-conditions' are replaced with a more general notion of the *propositional content* of an assertion (or stative sentential attitude), representing the *increment of information* it conveys (or stores): propositional content is connected to semantic value by (a) positing an array of *contexts* (representing concrete situations of speech or reasoning), and (b) requiring that an assertion using a certain sentence φ taking place in a context c (or an attitude held toward φ by the inhabitant of a context c) be assigned a proposition $\varphi(c)$ determined by c and the semantic value of φ[2];

S3T. Logical consequence remains *truth-preservation*, with 'logical truth-conditions' analyzed by relating contexts and propositions to *possible worlds*:

(a) Contexts are interpreted as 'possible locations' and represented as *centered possible worlds*, with a context c severally determining and collectively determined by a possible world w_c, moment of time t_c, and individual agent a_c;

(b) Propositions are analyzed as (determining) *possible-worlds truth-conditions*;

The sentence φ is then *true in a context c* just if the proposition $\varphi(c)$ is true in the contextually actual world w_c—so that premisses Ψ entail a conclusion φ just if for every context c: if (for every $\psi \in \Psi$, $\psi(c)$ is true in w_c), then $\varphi(c)$ is true in w_c.

Return now to the various semantic gaps: by (II), each involves a true mental sentence $\psi^\dagger$ such that for every true physical sentence φ, endorsing φ without $\psi^\dagger$ is intelligible; so, by (I), this $\psi^\dagger$ is not entailed by any such φ; so, by (S3T), for this $\psi^\dagger$ and any such φ, there is some context $c^\dagger$ such that $\varphi(c^\dagger)$ is true in $w_{c^\dagger}$ but $\psi^\dagger(c^\dagger)$ is false in $w_{c^\dagger}$. But then the contextual truth-value of $\psi^\dagger$ is underdetermined by that of any physical truth φ (after all, both φ and $\psi^\dagger$ are *true* (recall), providing a context—namely, *our* context, c^*—such that both $\varphi(c^*)$ and $\psi^\dagger(c^*)$ are true in c^*).[3]

---

[2]The initial case for pragmatical truth-conditions shows that this determination cannot be compositional: rather, contribution from a 'postsemantic' stage is required.

[3]A consequence of (S3T) and right-to-left (I) and (II-C) is (a certain interpretation of) Chalmers's *CP*: 'if ψ is conceivable, ψ is 1-possible' (Chalmers 2010b, 166, renotated). Let ψ be such that (*) some φ does not transmit supposition to not-ψ; then by (II-C), some φ does not transmit endorsement to not-ψ; then by (I), some φ does not entail not-ψ; then by (S3T), ψ is true in some context. CP follows, interpreting 'conceivability' with (*) and '1-possibility' with truth in some context.

The contextual truth-value underdetermination of some mental truth by any physical truth yields a progression of *dispiriting consequences*:

- *Nonphysicalist metaphysics*: both the mental truth $\psi^\dagger$ and every physical truth φ are, of course, true in our context c^*: but a context $c^\dagger$ with every such φ still true but $\psi^\dagger$ false is perhaps (Chalmers 2010b, 146) a possibility that *we* be such that these φ are true but $\psi^\dagger$ is false—so at least how *we are* physically underdetermines how *we are* mentally.
  Say that a sentence is *intraworld-insensitive* just if whenever contexts c and c' share an actual world, they assign the sentence the same truth-value; otherwise *intraworld-sensitive*. Perhaps such intraworld-sensitivity in $\psi^\dagger$ is not required by the semantic gap (Chalmers 2010b, 163); perhaps the gap survives its explicit suppression (Chalmers 2010b, 162). If so, $\psi^\dagger$ is false in the world $w^\dagger$ of $c^\dagger$ but true in the world @ of our context c^*, despite the truth in both $w^\dagger$ and @ of every intraworld-insensitive physical truth φ: so how *the world is* physically underdetermines how *the world is* mentally.
  Last, say that a sentence φ is *context-insensitive* just if for any contexts c and c', $\varphi(c) = \varphi(c')$. Perhaps context-sensitivity in $\psi^\dagger$ is not required by the semantic gap: then there is some proposition $q^\dagger$ such that invariably $q^\dagger = \psi^\dagger(c)$; and for every proposition p which is the content of some true context-insensitive physical sentence, the possible-worlds truth-value of p underdetermines that of $q^\dagger$ (Chalmers 2010b, 149).[4] On a familiar interpretation (Kripke 1980, 153–4), at least some mental facts are 'superadded' to the physical facts—to 'make the world', God did not stop after making its physical aspect.
- *Epiphenomenalism*: if nonphysicalism is discomfiting already,[5] worse is to follow. If certain mental facts are even *apparently* superadded to the physical facts,[6] this undercuts the legitimacy of 'diachronic' explanations running between such facts and the physical facts. Many of us are convinced that physics is 'causally closed' (compare Lewis 1966, 105): when God made the physical aspect of the world, this completed the dossier of facts about the causal impingements on (and perhaps *by*) the physical aspect of the world; if there was a subsequent mental superaddition, it involved no further contribution of causal impingements on (or perhaps *by*) the physical. But this requires denying any superadded mental facts the power to causally impinge on (or perhaps be impinged upon by) the physical.

[4]To save space, yet further complications from the prospect of inevitable context-sensitivity are suppressed: compare Chalmers (2010b, 150–52)—but the 'Russellian monism' emerging here is vulnerable to 'conceivability of the mind without the body' (Nagel 1970, 401, repeating from above).

[5]Compare: 'the antecedent impulse to believe materialism [is] so strong (I share it, too), and my conclusions so hard to accept' (Chalmers 1999, 3.6).

[6]Perhaps, contrary to 'modal rationalism' (Chalmers 2010b, section 10), the underdetermination goes for 'merely epistemic' possible worlds: if this blocks the case against physicalism, the current worry yet remains intact.

Unfortunately, in merely forming an intention to set ourselves a reminder alarm to phone Mom, we seem to 'presuppose'[7] that forming the intention (mental) will be somehow responsible for a later setting of the alarm (physical), which will in turn be somehow responsible for a still later forming of an intention to phone Mom (mental). If this 'responsibility' is causal, our presupposition is incompatible with the assumed superaddition of the involved mental facts. 'Noncausal responsibility', though, is hard to understand (Davidson 1963)[8]; and yet without it, we may be torn between *practical* reason (which demands we presuppose mutual responsibility between mental and physical facts) and *theoretical* reason (which demands we not contradict ourselves).

- *Separatism*: over the 1970s, popularity massed behind a strategy of avoiding contradiction by separating a 'responsible but gapless' *functional* understanding of the mental, from a 'gappy but irresponsible' *phenomenal* understanding (compare Shoemaker 1975; Jackson 1977; and especially Fodor 1991, 12 on 'dividing and conquering').

  Unfortunately, this separatism seems unlikely to get off the ground: my ordinary, first-person understanding of my own mental life is simultaneously *both* gap-creating and used in presupposing responsibility.[9] Perhaps we must choose between practical and theoretical reason, after all!

17.3 Endorsement-Conditionalism

Fortunately, the choice is not forced, but rather a creature of theory—specifically, of protosemantics. We may therefore avoid it through protosemantic revision: abandoning *truth* as the fundamental relation of meaning, in favor of *endorsement*.

[7]Compare Fodor (1989, 77): 'If it isn't literally true that my wanting is causally responsible for my reaching, and my itching is causally responsible for my scratching, and my believing is causally responsible for my saying ... if none of that is literally true, then practically everything I believe about anything is false and it's the end of the world'.

[8]Particularly so, for truth-conditionalists; less so for endorsement-conditionalists—as per below (and Hellie 2018).

[9]The complaint fast-marches a slowly evolving dialectic from the literature (for details, see Hellie 2019). Toward the end of the 1980s, the elusiveness of any nonexplanatory 'phenomenal' understanding would provoke widespread dissatisfaction (paradigmatically, the concept *red* applies to external surfaces; the concept *looks red*, if mental, has explanatory power, making it 'functional'; the concept *phenomenal red*, if distinct from these others, does not apply to anything easily identified: compare Dennett 1988; Lewis 1988b), mounting to rebellion by the early 1990s (Harman 1990).

By the mid-2000s, *nonreductive representationalism* (*phenomenal red* is a distinctively 'phenomenal' sense of *looks red*: compare Chalmers 2004b) had emerged to soothe the elusiveness worry: apparently, by subdividing the 'functional' into the 'representational' and the 'explanatory' and folding the 'phenomenal' in with the former: the presumptive difficulty next in line, the elusiveness of any nonexplanatory 'phenomenal-representational' understanding, has so far itself eluded widespread attention.

This marks a drastic recentering of the theory of meaning, from the world to the mind: whether 'goats eat cans' is *true* is constant among subjects sharing the same actual world, and varies among worlds only in accord with how things are in those worlds; whether 'goats eat cans' is *endorsed* varies from individual subject to individual subject (even subjects sharing the same actual world), in accord with whether their mental state is apt to endorse it. For truth-conditionalism, the meaning-giving condition for 'goats eat cans' targets a single entity, common to all (actual-world) subjects, and is satisfied or not regardless of how anyone is mentally; for endorsement-conditionalism, that condition targets a distinct entity for each subject at each time, and is satisfied or not exactly as a matter of how that subject is then mentally.

Ranked by first-glance plausibility, endorsement-conditionalism is not below truth-conditionalism, but arguably above: nebulously if perhaps plausibly, the meaning of a sentence is what it is used to mean—namely, as an archetype, *one's occupying a mental state sufficient for its endorsement*, as displayed through its use in assertion (for a declarative-mood sentence; otherwise, in that distinct mood-appropriate archetypal speech act). Moreover, truth-conditionalism is implausibly stringent, endorsement-conditionalism credibly flexible, regarding whether every meaningful sentence must in some sense have a truth-condition: for an interrogative ('who shot JR?') or imperative ('pay the rent!') sentence, no such sense is evident; by contrast, such sentences plausibly do have endorsement-conditions—perhaps (respectively) *wondering who shot JR* and *intending (salient) so-and-so to pay the rent*. Moreover, when assignment of a truth-condition is plausible ('goats eat cans'), so is assignment of an endorsement-condition (*holding an information-state endorsing the proposition that goats eat cans*): perhaps, of course, when a mental state meets the endorsement-condition, that mental state itself, by virtue of the propositions endorsed in its information-state, has a truth-condition; perhaps then the sentence inherits that truth-condition 'by courtesy'—but the move does not generalize beyond the declarative, in the absence of any evident sense in which *wondering who shot JR* or *intending so-and-so to pay the rent* imposes a truth-condition on the world.

The attitudes of *wondering* and *intending* are not alone in their failure to impose any condition on the world for their truth: joining them is *supposition* in various varieties (and under various labels: *hypothesis*; *pretense*; *purport*)—which we might analyze as a 'subsidiary' mental state, held (perhaps alongside any number of other suppositions) within one's 'root' mental state, and potentially bearing any aspect available in a 'root' mental state (including 'daughter' suppositions of its own). Holding a supposition that goats eat cans—a supposition whose information-state represents that goats eat cans—does not thereby impose any truth-condition on the world: even if one *suppositionally* imposes the truth-condition that the world be *such that goats eat cans*, to do something suppositionally is not thereby to do it outright. So if a sentence with a *wondering*- or *intending*-based endorsement-condition lacks truth-conditions (even by courtesy), the same would be true of a sentence with a *suppositional* endorsement-condition—a condition satisfied just by those subjects holding an appropriate sort of supposition.

Are there any? A first-glance plausible case is the *metafictional* sentence, like 'in 'The adventure of the speckled band', a Russell's viper climbs a rope'. Perhaps endorsing it requires holding a supposition that a Russell's viper climbs a rope, in a distinctive manner involving 'The adventure of the speckled band'—more specifically, perhaps, by committing the supposition to hold such-and-such information whenever the text of 'Speckled band' asserts such-and-such. The account is intuitively plausible, and moreover does not as its first move postulate a novel ontology (crucially, for those seeking a theory of fiction unencumbered by metaphysical perplexities).

Suppositional endorsement-conditions for *mental* sentences are not far off. According to *simulationists* (Heal 2003), metapsychological reasoning makes essential use of empathy with the other. Empathizing with Fred is 'occupying Fred's viewpoint': not *literally*, of course, but *suppositionally*—by holding a 'Fred-simulation': a supposition held in a distinctively Fred-viewpoint-occupying manner. To sketch some further details:

Σ1. For a supposition to count as 'occupying *a* viewpoint', it should be constrained by 'reasonability'; for the viewpoint to count as *Fred's* (at a certain time), it should be constrained empirically by information about Fred (then).
Sharpening, 'reasonability' can be understood as *similarity to oneself* (Heal 1998); the information plausibly concerns Fred's (wide) *sensory stimulations* and (wide) *behavioral responses* (and perhaps nothing else: Lewis 1974).
Sharpening more, we may allow that, in a given mental state, one finds some mental states easier to occupy, others harder: this can be represented with an *availability*-ordering on mental states (easier occupation goes with greater availability), with one's own mental state uniquely maximally available; and we may require that a mental state is *Fred-apt* by certain information only if its 'evidence of the senses' has content entailed by the facts of Fred's sensory stimulations (by the information) and its 'intentions-in-action' have content entailed by the kinds of Fred's behavioral responses (by the information).
Assembling, one's simulation of Fred is that mental state *most available* (to one) among those mental states which are *Fred-apt* (for one).[10]

Σ2. When 'the other' is *oneself*, the aptness-constraint is null: one's 'self-simulation' is just the mental state most available from one's own, *simpliciter*—namely, one's own mental state. (So one simulates oneself as Ψ-ing just if one Ψs, and as not Ψ-ing just if one does not Ψ: akin to 'Moore-paradoxicality'.)

Σ3. Empathy is sometimes at a *total loss*: facing the prospect of occupying the point of view of this table, I find that I cannot: by my information, *no* available mental state is 'this table-apt'.[11]

---

[10]The format is structurally akin to the Stalnaker conditional (Stalnaker 1968), where 'if ψ' shifts a world to its most similar ψ-world.

[11]Moreover—contrasting with (Σ5c)—this unavailability is 'robust', persisting as my information is weakened.

Σ4. Empathy is other times at a *partial loss*. Consider (prerelease) Black-and-White Mary: some things about what it will be like for her upon first seeing a red thing, she knows (that she will then recognize herself to see a color previously unseen and unimaginable; that she will then feel excited at this); others, she famously does not know, and cannot get to know by reckoning (that—as I would put it—she will then recognize herself to see red qua *this color*).

Mary's ignorance can be traced to an imprecision in how she simulates her future self: while we now can simulate post-release Mary precisely (in all relevant respects), and post-release Mary will come also to be able to simulate herself, Mary's pre-release condition fails to single out our way of simulating, as against others she has yet to eliminate. In consequence, pre-release Mary discriminates states involving recognizing oneself to see a color previously unseen from states not involving this; but she does not discriminate states involving recognizing oneself to see *red* qua *this color* from those involving recognizing oneself to see *green* (or *blue*, and so forth) qua *this color*. Failing to discriminate these latter, she cannot even entertain them individually in thought for sake of supposition, let alone simulate someone (including her future self) as occupying a specific one of them.

To generalize, we adjust the definition of *one's simulation of Fred*: it is that set of mental states, each most available among the Fred-apt, according to some way of simulating one has failed to eliminate Fred as one simulates him is just those ways common to all states in the set.

Σ5. Other complications handle uncertainty from other sources: (a) *partial information* (several x-apt states or discriminable regions are tied for most available); (b) *imperfect intelligibility* (a state/region is *imperfectly x-apt* just if apt to some weakening of one's information about x: then x is *imperfectly intelligible* when no x-apt states/regions are available, but several imperfectly x-apt states/regions are: compare Lewis 1982)—across the board, the way Fred is simulated to be is what is common to the several simulations.

Σ6. Mental–physical explanations are distinctive in that they involve a trip through supposition (compare Hellie 2018). Consider an example dialogue:

> A: Why did Fred arrive at noon [physical]? B: Because several days ago he resolved to set an alarm to get him there by then [mental]. A: But granting that, why? B(1): Because Fred resolved to set an alarm, he did [physical]. A: But granting that, why? B(2): Because Fred set the alarm, it went off [physical]. A: But granting that, why? B(3): Because the alarm went off, Fred noticed it [mental]. A: But granting that, why? B(4): Because Fred noticed the alarm, he set about acting to arrive at noon [mental]. A: But granting that, why? B(5): Because Fred set about acting to arrive at noon, he did [physical].

Explanation (1) is mental-to-physical: supposing oneself (qua Fred several days ago) to resolve to set an alarm, the supposition is more reasonable (more 'available') if it includes evidence of the senses and intentional actions characteristic of setting an alarm; but then the *aptness* constraint requires information about Fred's sensory stimulation and behavioral responses to the effect that the alarm is set. Explanation (2) is physical-to-physical, and evolves the information that the alarm has been set in conformity to expectations, yielding the information that the alarm went off. Explanation (3) is physical-to-mental: with the information that the alarm went off near Fred some time before noon and was therefore among his sensory stimulations then, the 'aptness' constraint (under the expectation that Fred notices the alarm) requires a supposition for Fred then with evidence that the alarm has gone off. Explanation (4) is mental-to-mental: carrying forward the intention with which the alarm was set from the Fred-several-days-ago supposition to the Fred-before-noon supposition, that latter supposition is then more available if it sets about acting to arrive at noon. Explanation (5) is mental-to-physical, returning to the pattern of explanation (1).

Explanation (2), the physical-to-physical portion, goes on just inside of the information-state, and involves no detour through the supposition: this, perhaps, is what is characteristic of 'causal' explanation. Explanation (4), the mental-to-mental portion, goes on just inside of the supposition, and involves no passage through the information-state: this, perhaps, is what is characteristic of 'rationalizing' explanation. But explanations (1) and (3) pass (respectively) from the supposition to the information-state and from the information-state to the supposition: this makes them neither (purely) 'causal' nor (purely) 'rationalizing'. Instead, they involve an essential *shift of view*, between the 'objective' viewpoint of causality and the 'subjective' viewpoint of rationality: the requirement of *aptness* constrains this shift of view, binding information about sensation and behavior to supposition about evidence and intentional action.

Armed with this conception of simulations (or perhaps some alternative), an endorsement-conditionalist could supply suppositional endorsement-conditions along the following lines: 'Fred intends to pay the rent' is endorsed by just those holding simulations of Fred as intending to pay the rent.

One might anticipate the sort of payoff that comes from moving burdens off of the world and on to the mind—rightly so, as will be discussed. But one might also fear that it is too good to be true: the approach is broadly 'expressivist'—and haven't all such attempts been shot down by the 'Frege-Geach problem', of handling 'embedded' occurrences (Geach 1960; Schroeder 2008)? Were we to impose a restriction to early-Davidson resources (deriving 'E-sentences' from axiom schemata using weak rules), *extreme skepticism* would be warranted: we might want the E-sentence *'Brent believes that Rance does not intend to pay the rent' is endorsed in c just if c simulates Brent as simulating Rance as not intending to pay the rent*—but how on earth to get it?

Still, goods are for goose and gander alike. The challenges of *now* and *actually* are, in effect, a Frege-Geach problem for early Davidson-style truth-conditionalism: resolving it requires the added articulation of the Standard Framework. Adapting the Standard Framework for endorsement-conditionalism requires only a minor tweak, of replacing (S3T) with (S3E):

S3E. Logical consequence is *endorsement-preservation* (compare Humberstone 1981; Yalcin 2007; Hellie 2016), analyzed by relating contexts and propositions to *information-states*:

(a) Contexts are interpreted as mental states; a context c determines at least a propositional information-state b_c (and also a time t_c and agent a_c: I would argue, if the mental state is that of subject a at time t, then $a = a_c$ and $t = t_c$);

(b) Propositions are analyzed as (determining) *endorsement-conditions* on *information-states*—more to follow shortly;

The sentence φ is then *endorsed in a context c* just if the proposition $\varphi(c)$ is endorsed in the contextual information-state b_c—so that premisses Ψ entail a conclusion φ just if for every context c: if (for every $\psi \in \Psi$, $\psi(c)$ is endorsed in b_c), then $\varphi(c)$ is endorsed in b_c.

This framework allows for drawing the needed distinction, between sentences with a broadly 'descriptive' meaning, involving an *informational* endorsement-condition, and sentences with instead a broadly 'expressive' meaning—including those with *suppositional* endorsement-conditions.

To help with this, we will sketch an endorsement-conditionalist interpretation of *propositions*. For truth-conditionalists, a proposition is (or determines) a truth-condition on possible worlds; do endorsement-conditionalists say a proposition is (or determines) an endorsement-condition on information-states? Sort of—but because information-states are propositions, this does not illuminate what either are like.

Fortunately, propositions are nailed down in (S2) by their role as *increments of information*. An increment of information, perhaps, essentially *answers questions*—under a modestly technical notion of 'question' (compare Hamblin 1963; Lewis 1988a; Groenendijk and Stokhof 1996):

Let a *polar* (yes/no) question be *objective* just if (i) both answers present intelligible ways for things to be (excluding, say, *is* $2 + 2 = 5$*?*: the *yes* answer is not intelligible), and (ii) each answer 'appears the same' from every vantage point (excluding, say, *is that a goat?*: the *yes* answer goes with *that is a goat*, which changes its appearance with what is being demonstrated).

For a pair of objective polar questions, there are four combinations of 'yes' and 'no' answers; some of these four (at least two) are intelligible ways for things to be: these are the 'total answers' to that unique question which is the *product* of that pair of questions—mutatis mutandis, for any (perhaps any *finite*) such set.

Last, a ('technical') *question* is the product of some set (perhaps some *finite* set) of objective polar questions; for a given question, its *answers* are the non-empty sets of its total answers; and an *answer* is an answer to some question or other.

Identifying increments of information with answers and propositions with increments of information, propositions are identified with answers: *trivial* and *absurd* propositions are limiting cases, identified with maximally weak and maximally strong increments of information; other propositions are *informative*.

Perhaps there is a set of all questions; and perhaps the product over this set is itself a question: if so, this latter is the *total question*, and its total answers are the *tototal answers*. Tototal answers and possible worlds appear quite similar: perhaps theory need not discriminate them. If so, then propositions qua answers qua sets of tototal answers on the one hand, and propositions qua sets of possible worlds on the other, are indiscriminable. So: the possible-worlds theory and the answer theory of propositions are indiscriminable, if there is a total question (for friends of the answer theory, this shows that the determination of a possible-worlds truth-condition is inessential to propositions, but at best a lucky accident). We embrace the total question to simplify explication of the sense in which a proposition p determines an endorsement-condition on (propositions qua) information-states: p determines a set S of tototal answers; any (proposition qua) information-state determines some set T of tototal answers, and endorses p just if T is a subset of S (namely, just if every tototal answer taken seriously in the information-state is compatible with p; just if the information state has ruled out any way of siding with not-p).

We can now distinguish 'descriptive' from 'expressive' sentences. When φ is *descriptive*, the proposition $\varphi(c)$ is (typically or always) informative, and the 'function' of φ is for its assertion in a context c is to convey that informative proposition $\varphi(c)$. An *expressive* sentence, by contrast, has a different 'function': one never asserts it to convey an informative proposition, but rather to display that one's mental state meets its endorsement-condition. Suppose, for example, that φ has a suppositional endorsement-condition—that endorsing φ requires nothing of a context/mental state c beyond c's holding such-and-such a supposition; the specific information carried in the information-state b_c is irrelevant. This requires that $\varphi(c)$ is such as to be endorsed (whatever the information-state of c) whenever c holds such-and-such supposition, and that $\varphi(c)$ is such as to not be endorsed (whatever the information-state of c) whenever c fails to hold such-and-such supposition. These in turn are implemented by setting the value of $\varphi(c)$ to the trivial proposition (endorsed whatever b_c may be), just if c holds the φ-appropriate supposition; and otherwise to the absurd proposition (rejected unless b_c is itself absurd)—a sort of pattern sometimes known as *testing the context* (compare Veltman 1996; Gillies 2004); in the present case, testing it for such-and-such supposition.

Detailed issues of how to combine this 'test of context' behavior with various embedding phenomena are beyond the scope of this discussion; as a 'proof of concept', however, the attached footnote briefly sketches a simulationism-friendly compositional semantic value for sentences ascribing beliefs.[12]

[12]Double verticals are semantic-value brackets (with $\|\varphi\|$ mapping a mental state, various indices, and a context to a proposition); for mental state m, $\Sigma_\tau(m)$ is the $\|\tau\|$-apt mental state most available from m; $m \Vdash p :=$ m's information-state endorses p; and $\top$ and $\bot$ the *trivial* and *absurd* propositions. Then: $\|\mathsf{B}^\tau\varphi\|(m, x, c) = \top$, or $\bot$, *just as whether* $\Sigma_\tau(m) \Vdash \|\varphi\|(\Sigma_\tau(m), x, c)$. Last (where m_c and x_c are the mental state and indices determined by context c), $\varphi(c) = \|\varphi\|(m_c, x_c, c)$. Combining 'tests' with uncertainty (as with Black-and-White Mary) requires, moreover, supervaluating over elements of context at the 'postsemantic' stage in which content is determined.

17.4 Semantic Gaps Within Endorsement-Conditionalism

Endorsement-conditionalism offers a fresh look at the semantic gap phenomena. Where these arise (as we have seen), they lead, via (I), (II), and (S3T) to the *contextual truth-value underdetermination* of the involved mental truths by any physical truth, and thence to the *dispiriting consequences*. But (S3T) is a specifically truth-conditionalist commitment, replaced by the endorsement-conditionalist with (S3E).

This concluding section *first* explains (swiftly) that semantic gap phenomena, (I), (II), and (S3E) do not lead to *contextual truth-value underdetermination*—nor therefore to the *dispiriting consequences*; and *second* sketches (more gradually) an endorsement-conditionalist analysis of the various semantic gap phenomena (highlights: the overwhelmingly plausible *epistemic gap* is legitimated; while truth-conditionalism assimilates to it the much more contentious *suppositional gap*, this assimilation is an artifact of the 'bivalence' of *truth*: by contrast, with the 'trivalence' of *endorsement*, the two gaps can be segregated, and due allowance given to their contrasting plausibility; finally, the *explanatory gap* is yet a third effect, emanating from the 'viewpoint shift' between physical and mental reasoning).

First: consider a worst case scenario: for *every* (consistent) mental sentence ψ, it is implicated in one or other sort of semantic gap with *every* (consistent) physical sentence φ: by (II), for any such φ and ψ, endorsing φ does not require endorsing either ψ or $\neg\psi$; by (I), for any such φ and ψ, φ entails neither ψ nor $\neg\psi$. But by (S3E), this just reiterates the endorsement-gap (if more formally): for any mental ψ and physical φ, there are contexts c' and c'' such that the information-state $b_{c'}$ endorses the proposition $\varphi(c')$ but not the proposition $\psi(c')$, while the information-state $b_{c''}$ endorses the proposition $\varphi(c'')$ but not the proposition $\neg\psi(c'')$.

The previous section sketches an analysis of mental sentences as *tests of context*: for any context/mental state c and mental sentence ψ, the proposition $\psi(c)$ is either *trivial* or *absurd*, just as whether c holds a ψ-appropriate simulation. So if the information-state $b_{c'}$ fails to endorse the proposition $\psi(c')$, $\psi(c')$ is absurd, and c' *does not* hold a ψ-appropriate simulation; and if the information-state $b_{c''}$ fails to endorse the proposition $\neg\psi(c'')$, $\neg\psi(c'')$ is absurd, $\psi(c'')$ is trivial, and c'' *does* hold a ψ-appropriate simulation. We may, moreover, pick any physical sentence φ we like: to maximize the challenge, let φ be context-insensitive and total—there is some maximally-specific proposition p such that for any context c, $p = \varphi(c)$. Then because both $b_{c'}$ and $b_{c''}$ endorse this proposition p, endorsement of this proposition is compatible both with holding the ψ-appropriate simulation and with failing to do so—for any increment of physical information and any simulation, holding the former neither requires nor prohibits holding the latter.

Still, the dispiriting consequences are neutralized:

- When c', but not c'', refrains from ψ-appropriate simulation, the dispute does not involve conflicting information, and is indeed compatible with *identifying* their information-states: whatever either thinks God did in making the world, the other agrees—their dispute is over the unrelated question of how each should

react empathetically to this or that agreed-upon aspect of reality: a conflict of 'sentiment', rather than over anything 'objective'.
Further dispute over *physicalist metaphysics* may well yet be pursued, but its forum will have to be outside of the philosophy of mind.

- We may grant that the physical is causally closed, and that one's physical information 'freely crosscuts' which simulations one holds, without withdrawing the legitimacy of appeals to mental–physical interaction. The discussion under (Σ6) illustrates how such appeals involve a mix of causal- and noncausal-explanatory reasoning. While free-crosscut would make for potentially irresoluble dispute over any particular candidate noncausal explanans, it would also license one to ignore such dispute and forge ahead with such appeals as are appropriate to one's sentiments.
The threat of *epiphenomenalism* is neutralized, by rising to the challenge from Davidson (1963), and offering an account of noncausal explanation.
- With epiphenomenalism out of the way, there is no longer any need to search around for a semantic gap *internal* to mental language: *separatism*, and the quest for 'phenomenality' it requires, is no longer useful in theory.

Second (now less urgently, having disarmed the *dispiriting consequences*): what to make of the semantic gap phenomena—which are genuine, to what extent, and in what do they consist?

Start with the *epistemic gap*: the epistemic gap judgment regarding Black-and-White Mary (and in related cases, such as bats, or the smell of skunks or taste of Vegemite) is overwhelmingly plausible: we should use the apparatus to analyze it.

Suppose that (prerelease) Mary has total physical knowledge about her future self upon first seeing a red thing (total physical knowledge *simpliciter*, if needed): this is captured in some true physical sentence φ. Let ψ be a truth along the lines of 'post-release Mary Ψs', where Ψ is a mental predicate discriminating more finely than Mary's resolution of the space of mental states—perhaps Ψ is *have as evidence that one foveates a red thing*; if not, the reader may coin a predicate to their liking. Plausibly, Mary does not know ψ: after all, as discussed under (Σ4), pre-release Mary does not fully grasp how to simulate her post-release self: if, as urged by Harman (1990), our grasp of 'red' is inextricable from our ability to simulate seeing red, pre-release Mary does not fully understand what the predicate means, and cannot endorse or know sentences using it. So with Mary knowing and endorsing φ but neither knowing nor endorsing ψ, φ does not entail ψ.

This failure of entailment does not, however, predict the consistency of φ with $\neg\psi$. Endorsement is *trivalent*: when premisses do not entail a conclusion, one way for this is for some context to endorse the premisses and also the negation of the conclusion; but another way is for some context to endorse the premisses but neither the conclusion nor its negation. (This contrasts with truth, which is *bivalent*: the only way for premisses not to entail a conclusion is for there to be a context verifying the premisses but falsifying the conclusion.) The stronger condition would require a subject who knows φ but who also endorses $\neg\psi$. But this sort of subject *disagrees* with our view of what it will be like for post-release Mary: this is very different from *failure to endorse* our view—and it is only the latter which is legitimated by the many widely-discussed examples of *failure to know what it is like*.[13]

Now to the *suppositional gap*. It is striking that the *epistemic* gap is scarcely controversial (and the Nemirow-Lewis 'ability hypothesis' (Nemirow 1980; Lewis 1988b) that Black-and-White Mary undergoes no 'rational-psychological' change widely derided), while the *suppositional* gap is significantly more so. We would like to explain the asymmetry, not just to settle whether 'zombies', or 'inverts', are 'conceivable'.

Observe at the outset that supposing φ *and Fred* Ψs involves 'nesting' suppositions: the 'outer' supposition is populated *both* with an information-state endorsing φ, *and* with an 'inner' supposition, which is a simulation of Fred as Ψ-ing. Presumably a nested simulation is still a simulation, and therefore still is constrained ($\Sigma 1$) to maximize availability subject to Fred-aptness: but the information-state to which *aptness* is pegged is not one's 'root' information, but that of the 'outer' supposition; whereas *availability* is pegged instead to the 'root' mental state, rather than the 'outer' supposition (otherwise, the standards of the 'outer' supposition for what constitutes reasonability would be involved: this would seem to be appropriate to, say, 'first-person fiction' about someone with information φ who thinks Fred Ψs; but what is wanted instead is just *my* supposition both that φ and that Fred Ψs).

So suppose we think Fred, a normal subject under normal conditions, foveates a red tomato: in consequence, we simulate Fred with a mental state such that what it is like to be in it is as if foveating a red tomato (here and henceforth leaving tacit 'ordinariness')—in particular, this is *not* a mental state such that what it is like to be in it is as if foveating a *green* tomato. And let us introduce a supposition in which things are physically exactly as they actually are. This, presumably, is exactly recapitulating my 'root' physical information in the suppositional information—and, if need be, amplifying it to complete physical information in a way that excludes any surprises (for instance, some hitherto unnoticed visual abnormality in Fred). Assume, to simplify, that all information is physical information[14]: then the suppositional information is *exactly* my root information.

But this starkly constrains the suppositional simulation of Fred. Because my 'root' and 'outer-supposition' information-states are the same, a mental state is 'suppositionally Fred-apt' (Fred-apt by the lights of the 'outer supposition' information-state) just if 'root Fred-apt' (Fred-apt by the lights of the 'root' information-state). And, recall, 'suppositional availability' is just 'root availabil-

[13]Compare Nagel (1974, 442n8): 'My point[] is not that we cannot *know* what it is like to be a bat. I am not raising that epistemological problem. My point is rather that even to form a *conception* of what it is like to be a bat [] one must take up the bat's point of view'. Nagel himself therefore advances a 'no concept' treatment of ignorance of what it is like: compare Harman (1990) and Hellie (2004); contrast Chalmers (2004a, 284): treating objections there, see just below on 'appeal to the conceivability of zombies'; ($\Sigma 4$), again, on '*other* organisms (bats or Martians[])' and perhaps 'phenomenal indistinguishability'; and ($\Sigma 2$) on 'introspection yields aposteriori knowledge'.

[14]See Lewis (1988b, 281) for why the assumption is innocent.

ity'. So the *suppositional* simulation of Fred is the most root-available of the suppositionally-Fred-apt mental states—namely, the most root-available of the *root*-Fred-apt mental states: so my *suppositional* and *root* simulations of Fred do not differ—so, in particular, the suppositional simulation is also a state such that what it is like to be in it is *not* as if foveating a green tomato.

But enriching the supposition so that Fred is inverted requires the suppositional simulation to be a mental state such that what it is like to be in it *is* as if foveating a green tomato. Unfortunately, this is incompatible with the supposition of physical identity. And so it cannot be coherently supposed that Fred is an 'invert' (similarly, because the suppositional simulation is such that what it is like to be in it *is* as if foveating a red tomato, there *is* a suppositional simulation—incompatibly with the supposition that Fred is a zombie); and so the request to suppose it is *unintelligible*.

Unintelligible to me, anyway. But *perhaps* someone else can conceive of inverts or zombies—*can*, intelligibly; intelligibly to *me*: perhaps *I* can make sense of someone's holding an invert or zombie supposition. Can I? Anyone who does, as argued, already thinks Fred is an invert or zombie; so the issue is whether I can make sense of someone—Zeb—who thinks this. Zeb shares my information, so the same mental states are Fred-apt for me as for Zeb; but among those, the most available for Zeb is either different (invert) or absent (zombie): an availability-structure very different from mine. If Zeb is intelligible, some Zeb-apt mental state is available from mine; if Zeb thinks Fred is an invert or zombie, that mental state has a very different availability-structure to mine. Whether this prospect is genuine is, at present, an open question.

The literature contains many confident proclamations that zombies/inverts do not exist, but are yet conceivable. As argued, these attitudes are incompatible, rendering those who so proclaim imperfectly intelligible: what they articulate cannot be had in mind, so what they have in mind has not been adequately articulated. How then, perhaps, to articulate what they really do have in mind? Several (jointly compatible) options come to mind:

(i) They imagine a system of pictures: circles, linked by arrows, some decorated with colors; clearly the nature of the pictorial system permits interchanging colors, or erasing them, leaving circles and arrows the same. The system of pictures is used as a calculus for reasoning about mental–physical relationships, presuming the separatist metapsychology: systems of circles and arrows inherit the meaning of 'functional' concepts; colors, of 'phenomenal' concepts. The permission from the nature of the pictorial system is treated as a permission within the calculus: the permission to interchange colors as a permission to conceive of inverts; to erase them, of zombies. The cost of revisiting separatism, or the adequacy of the pictorial system as a calculus even granting separatism, is high, the benefits unclear; the costs of its continued use are unclear, the benefits significant and known: efficiency favors continuing with 'normal science'.

(ii) They affirm the predictions of well-attested theory. After all, there is an evidently genuine *epistemic gap* between a certain mental $\psi^\dagger$ and any physical

φ; by (II) and (I), this yields the nonentailment by any φ of $\psi^\dagger$. By (S3T), this nonentailment requires some context verifying (for any φ) $\neg\psi^\dagger \wedge \varphi$; but by (S3T) again, this predicts the 'non-antivalidity' of (for any φ) $\neg\psi^\dagger \wedge \varphi$. This finally, by (I) and (II-C), yields the supposability of (for any φ) $\neg\psi^\dagger \wedge \varphi$—the conceivability of zombies or inverts.

(iii) They implicitly recognize the incoherence of their view, but cannot articulate its source. On the present account, the incoherence stems from intelligibility constraints on suppositional 'children' by 'parent' mental states. Our philosophical tradition has so far prioritized 'intralevel' coherence constraints (synchronically, on information-states; diachronically, among prior and posterior information-states and accumulated evidence; synchronically, between credence, value, and decision); for that matter, even taking note of the relevance of supposition requires 'simulationism', itself a minority doctrine. In consequence, the conceptual/descriptive resources required to characterize the incoherence are lacking; with innocence presumed and the nature of guilt unarticulated, one reasonably sides with conceivability.

Last to the *explanatory gap*. Like the epistemic gap, the supporting judgments (Leibniz's mill, for example) are highly plausible: we should again use the apparatus to analyze it.

To do so, it will be useful to map connections among several closely related phenomena:

(i) φ (constitutively) explains ψ
(ii) Granting φ forecloses wondering why ψ
(iii) Granting φ forecloses not accepting ψ
(iv) Granting φ forecloses rejecting ψ

It is hard to discern (i) from (ii): let them be equivalent. Trivalence blocks the way from (iii) to (iv) (avoiding the *dispiriting consequences*). Having stipulated equivalence of (i) and (ii), are these required by (iii)?

Perhaps not. Consider the *aptness* constraint on simulation. Bracketing the various sources of uncertainty under (Σ4) and (Σ5), let us assume a (perhaps for still other reasons, implausibly demanding) aptness constraint on which, with information that Fred's sensory stimulation is F, one must simulate Fred as having evidence that his sensory stimulation is F. By this constraint, granting that Fred's sensory stimulation is F *does* foreclose leaving open whether Fred has evidence that his sensory stimulation is F.

Now, presumably we should deny every instance of (i) and of (ii), for physical φ and mental ψ—how could an 'objective' fact constitute a 'projection of sentiment'? One such instance to be denied is: granting that Fred's sensory stimulation is F *forecloses wondering why* Fred has evidence that his sensory stimulation is F. But we just accepted the corresponding instance of (iii)—so there is more to φ *explaining* ψ than φ *foreclosing not accepting* ψ (on the affirmative side).

The difference, it would seem, is that φ foreclosing not accepting ψ makes for *explanation* only when φ and ψ are appreciated from the same perspective:

in the case of physical–physical constitution (water and H_2O), this is from the 'objective' perspective; if there is mental–mental constitutive explanation (perhaps *belief that Fred sees red* is constituted by information and availability), this is from the 'subjective' perspective. By contrast, foreclosure by the aptness constraint essentially involves a viewpoint-shift between the objective and subjective view, and therefore does not rise to the level of explanation. In particular, without accommodating this, pursuit of the 'hard problem of consciousness' (Chalmers 1995) will continue to be fruitless.

References

Broad, C. D. (1925). *Mind and its place in nature*. London: Kegan Paul, Trench, Trubner & Co.

Chalmers, D. J. (1995). Facing up to the problem of consciousness. *Journal of Consciousness Studies, 2*, 200–219. Reprinted as chapter 1 of Chalmers (2010a).

Chalmers, D. J. (1999). Materialism and the metaphysics of modality. *Philosophy and Phenomenological Research, 59*, 473–496.

Chalmers, D. J. (2002a). Consciousness and its place in nature. In D. J. Chalmers (Ed.), *Philosophy of mind: Classical and contemporary readings*. Oxford: Oxford University Press.

Chalmers, D. J. (Ed.). (2002b). *Philosophy of mind: Classical and contemporary readings*. Oxford: Oxford University Press.

Chalmers, D. J. (2004a). Phenomenal concepts and the knowledge argument. In P. Ludlow, Y. Nagasawa, & D. Stoljar (Ed.), *There's something about Mary*. Cambridge, MA: The MIT Press.

Chalmers, D. J. (2004b). The representational character of experience. In B. Leiter (Ed.), *The future for philosophy*. Oxford: Oxford University Press. Reprinted as chapter 11 of Chalmers (2010a).

Chalmers, D. J. (2010a). *The character of consciousness*. New York: Oxford University Press.

Chalmers, D. J. (2010b). The two-dimensional argument against materialism. In Chalmers (2010a), chapter 6. Oxford: Oxford University Press.

Davidson, D. (1963). Actions, reasons, and causes. *Journal of Philosophy, 60*, 685–700. Reprinted in Davidson (1980).

Davidson, D. (1967). Truth and meaning. *Synthese, 17*, 304–323. Reprinted in Davidson (1984).

Davidson, D. (1980). *Essays on actions and events*. Oxford: Oxford University Press.

Davidson, D. (1984). *Inquiries into truth and interpretation*. Oxford: Oxford University Press.

Davidson, D., & Harman, G. (Eds.). (1972). *Semantics of natural language*. Dordrecht: D. Reidel.

Dennett, D. C. (1988). Quining qualia. In A. Marcel & E. Bisiach (Eds.), *Consciousness in contemporary science*. Oxford: Oxford University Press. Reprinted in Chalmers (2002b).

Descartes, R. (1637/1985). *Discourse on method* (Vol. 1). Cambridge: Cambridge University Press. In *The philosophical writings of Descartes*, edited by Cottingham, Stoothof, and Murdoch.

Fodor, J. A. (1989). Making mind matter more. *Philosophical Topics, 17*, 59–79.

Fodor, J. A. (1991). Too hard for our kind of mind? *London Review of Books, 13*, 12.

Geach, P. T. (1960). Ascriptivism. *The Philosophical Review, 69*, 221–225.

Gillies, A. S. (2004). Epistemic conditionals and conditional epistemics. *Noûs, 38*, 585–616.

Groenendijk, J., & Stokhof, M. (1996). Questions. In J. van Benthem & A. ter Meulen (Eds.), *Handbook of logic and language*. Amsterdam: Elsevier.

Hamblin, C. L. (1963). Questions aren't statements. *Philosophy of Science, 30*, 62–63.

Harman, G. (1990). The intrinsic quality of experience. In J. Tomberlin (Ed.), *Action theory and the philosophy of mind* (Volume 4 of philosophical perspectives, pp. 31–52). Atascadero: Ridgeview. Reprinted in Harman (1999).

Harman, G. (1999). *Reasoning, meaning, and mind.* Oxford: Oxford University Press.
Heal, J. (1998). Other minds, reason, and analogy. *Proceedings of the Aristotelian Society, Supplementary Volume, 74*, 477–98. Reprinted in Heal (2003).
Heal, J. (2003). *Mind, reason, and imagination.* Cambridge: Cambridge University Press.
Hellie, B. (2004). Inexpressible truths and the allure of the knowledge argument. In P. Ludlow, Y. Nagasawa, & D. Stoljar (Eds.), *There's something about Mary.* Cambridge, MA: The MIT Press.
Hellie, B. (2016). Obligation and aspect. *Inquiry, 58*, 398–449.
Hellie, B. (2018). Praxeology, imperatives, and shifts of view. In R. Stout (Ed.), *Procedure, action, and experience.* Oxford: Oxford University Press.
Hellie, B. (2019). An analytic-hermeneutic history of Consciousness. In K. M. Becker & I. Thompson (Eds.), *Companion to the history of philosophy, 1945–2016.* Cambridge: Cambridge University Press.
Humberstone, L. (1981). From worlds to possibilities. *Journal of Philosophical Logic, 10*, 313–339.
Jackson, F. (1977). *Perception: A representative theory.* Cambridge: Cambridge University Press.
Jackson, F. (1982). Epiphenomenal qualia. *Philosophical Quarterly, 32*, 127–36.
Kamp, H. (1971). Formal properties of 'now' . *Theoria, 37*, 227–274.
Kaplan, D. (1977). Demonstratives. In J. Almog, J. Perry, & H. Wettstein (Eds.), *Themes from Kaplan.* Oxford: Oxford University Press. Published 1989.
Kripke, S. A. (1980). *Naming and necessity.* Cambridge, MA: Harvard University Press. Reprinted from Davidson and Harman (1972), with new preface.
Leibniz, G. W. (1714/1991). *Monadology.* New York: Hackett.
Levine, J. (1983). Materialism and qualia: The explanatory gap. *Pacific Philosophical Quarterly, 64*, 354–361.
Lewis, D. (1966). An argument for the identity theory. *Journal of Philosophy, 63*, 17–25. Reprinted in Lewis (1983).
Lewis, D. (1970). General semantics. *Synthese, 22*, 18–67. Reprinted in Lewis (1983).
Lewis, D. (1974). Radical interpretation. *Synthese, 27*, 331–344. Reprinted in Lewis (1983).
Lewis, D. (1980). Index, context, and content. In S. Kanger and S. Öhman (Eds.), *Philosophy and grammar.* Dordrecht: Reidel. Reprinted in Lewis (1998).
Lewis, D. (1982). Logic for equivocators. *Noûs, 16*, 431–441. Reprinted in Lewis (1998).
Lewis, D. (1983). *Philosophical papers* (Vol. I). Oxford: Oxford University Press.
Lewis, D. (1988a). Statements partly about observation. *Philosophical Papers, 17*, 1–31. Reprinted in Lewis (1998).
Lewis, D. (1988b). What experience teaches. *Proceedings of the Russellian Society, 13*, 29–57. Reprinted in Lewis (1999).
Lewis, D. (1998). *Papers in philosophical logic.* Cambridge: Cambridge University Press.
Lewis, D. (1999). *Papers in metaphysics and epistemology.* Cambridge: Cambridge University Press.
Nagel, T. (1970). Armstrong on the mind. *The Philosophical Review, 79*, 394–403.
Nagel, T. (1974). What is it like to be a bat? *The Philosophical Review, 83*, 435–450.
Nemirow, L. (1980). Review of Nagel's *Mortal questions. The Philosophical Review, 89*, 473–477.
Schroeder, M. (2008). *Being for.* Oxford: Oxford University Press.
Shoemaker, S. (1975). Functionalism and qualia. *Philosophical Studies, 27*, 292–315.
Stalnaker, R. C. (1968). A theory of conditionals. In N. Rescher (Ed.), *Studies in logical theory.* Oxford: Blackwell.
Stalnaker, R. C. (1970). Pragmatics. *Synthese, 22.* Reprinted in Stalnaker (1999).
Stalnaker, R. C. (1978). Assertion. *Syntax and Semantics, 9*, 315–332. Reprinted in Stalnaker (1999).
Stalnaker, R. C. (1999). *Context and content.* Oxford: Oxford University Press.
Veltman, F. (1996). Defaults in update semantics. *Journal of Philosophical Logic, 25*, 221–261.
Yalcin, S. (2007). Epistemic modals. *Mind, 116*, 983–1026.

Chapter 18
The Observer and Access to Information in the Quantum Universe

Menas C. Kafatos and Ashok Narasimhan

Abstract We examine aspects of observation, measurement, the role of the observer, and information in the quantum framework of the universe. The results of a class of quantum eraser experiments carried out in the laboratory can be interpreted as indicating that the collapse of the wave function may not necessarily involve choices by observers in space-time. Instead, it is the access to and interpretation of information, outside of space and time, which may be involved. Although not contradicting the standard orthodox interpretation, we believe it extends it and enriches it. Our interpretation is consistent with the presence of a universal Observer in the quantum universe of possibilities. The implications for the quantum universe and the role of the mind are discussed and explored, such as the need for a universal Observer. There is a possibility that individual observers making choices in space and time are really aspects of the existence of the universal Observer.

18.1 Introduction

In the Copenhagen Interpretation proposed initially by N. Bohr, W. Heisenberg, and further developed by them, including W. Pauli, M. Born and others (Harrison 2002; Cassidy 2008), the role of observation plays a fundamental role in the so-

---

Quanta and Mind Conference, April 10–11, 2018, SFSU Main Campus.
We dedicate the present work to Akshay Narasimhan.

M. C. Kafatos (✉)
Fletcher Jones Endowed Professor of Computational Physics, Chapman University, Orange, CA, USA

Korean Academy of Science and Technology, Seongnam, South Korea
e-mail: kafatos@chapman.edu

A. Narasimhan
California Institute for Integral Studies, San Francisco, CA, USA

Omnyway Corporation, San Mateo, CA, USA

J. Acacio de Barros, C. Montemayor (eds.), *Quanta and Mind*, Synthese Library 414, https://doi.org/10.1007/978-3-030-21908-6_18

called collapse of the wave function. The traditional Copenhagen was revised and extended by John von Neumann (1995), and now often referred to as the Orthodox Interpretation (heretofore, and in agreement with Stapp (2007, 2009), we refer to it as standard quantum mechanics, QM). In this standard Orthodox QM, there are two main types of processes: Process 1, Process 2, the second being the predictable, linear time evolution of the wave function, through the Schrödinger equation; while the first presents non-linear aspects of QM, wherein a collapse takes place upon observation. In this Orthodox Interpretation of QM, the act of observation is *essential* for collapse of the wave function. What this implies is that out of a large set of possible outcomes, observation confers reality on the specific outcome of a measurement. However, which of the many possible outcomes contained in the wave function will materialize (Process 1) is not known. The implication is that a human being decides to observe the properties of a quantum system at a particular point in time and space, setting up appropriate experiment, as for example to determine the direction of the spin of a particle. Although the process can be automated, this does not take out the participation of an observer, in the words of Wheeler (1981) "no phenomenon is a phenomenon until it is an observed phenomenon". The mind that plans, executes and subsequently makes the observation, is the mind of the observer, in the same space-time coordinates.

It is generally assumed that human observers are essential, can be considered to be local observers. We denote in the rest of the present work, such (human) observers by lower case ***o*.** Observer ***o*** sets up the quantum apparatus needed to conduct the observation or 'probing action' to ask a question of Nature. Nature responds with its answer (Process 1) at some initial time. This answer is recorded by the measuring apparatus. In general, the observer then observes the outcome which could be in the form of recording of the answer at a later time. The act of observation causes the particular form of Nature's answer – e.g. particle or interference pattern. Here we note that this is based on the concept of a conscious observer using a traditional measuring apparatus to measure Nature's answer, all occurring at the initial time.

In the basic double slit experiment, a beam of light, often from a laser, is directed perpendicularly towards a wall which has two parallel slit apertures. If a detection screen is put on the other side of the double slit wall, an interference pattern of light and dark fringes will be observed. By decreasing the brightness of the source sufficiently, individual particles that form the interference pattern are detectable (Donati et al. 1973). A well-known thought experiment, which played a vital role in the history of quantum mechanics, demonstrated that if particle detectors are positioned at the slits, showing through which slit a photon goes, the interference pattern will disappear (Feynman et al. 1965). There is a question of whether the act of observation disturbs the system being observed such that the results cannot be trusted to accurately represent its original state, in other words, the interference pattern gets disturbed enough by the act of observation, as described by Heisenberg's Uncertainty Principle, that it collapses to a particle.

Several questions arise, "does measurement require a conscious observer"? Even if fully automated, it would seem that the probing questions asked of Nature, practically require a measurement to be designed and carried out in the laboratory. In Orthodox QM, it appears that the participation of an observer is required. Otherwise, how would information itself be relevant in quantum experiments? In what follows, we raise some of these questions and explore relevant issues.

18.2 Quantum Eraser Experiments

Narasimhan and Kafatos (2016) examined the complementary aspects of knowing which way the particle followed in the so-called delayed choice experiments. The "which-way" delayed choice experiments illustrate the complementarity principle, inthat photons can behave as either particles or waves, but not both at the same time (Harrison 2002; Cassidy 2008; Boscá Díaz-Pintado 2007). As pointed out by Narasimhan and Kafatos (2016), technically feasible realizations of this experiment were not proposed until the 1970s (Bartell 1980). As Narasimhan and Kafatos pointed out, which-path information and the visibility of interference fringes, are complementary aspects of the experimental set up. In the double-slit experiment, conventional wisdom held that observing the particles inevitably disturbed them enough to destroy the interference pattern as a result of the Heisenberg uncertainty principle.

As discussed in Narasimhan and Kafatos (2016), in 1982, Scully and Drühl further developed the delayed-choice set up, by finding a loophole around a standard delayed choice experiment and its interpretations (Scully and Drühl 1982). They proposed a "quantum eraser" set up to obtain which-path information without scattering the particles or otherwise introducing uncontrolled phase factors to them. In other words, rather than attempting to observe which photon was entering each slit, and in this way disturbing them, they proposed to "mark" photons with *information* that would allow photons to be distinguished after passing through the slits. As anticipated, the interference pattern does disappear when the photons are so marked. Yet, the interference pattern *reappears* if the which-path information is further manipulated "after" the marked photons have passed through the double slits to obscure the which-path markings. It is clear that "before" and "after" are classical concepts which break down in actual quantum experiments and the information they imply (Narasimhan and Kafatos 2016). The actual "eraser" situation has since 1982, been experimentally have been verified (Zajonc et al. 1991; Herzog et al. 1995; Walborn et al. 2002).

Narasimhan and Kafatos (2016) have pointed out that quantum eraser set ups using entangled photons, are intrinsically non-classical. The situation is illustrated in the Kim et al. (2000) setup, shown in Fig. 18.1. There we show the different "paths" giving the two complementary situations, the "which path" (revealing the particle aspect), versus the "scrambled" (revealing the wave aspect). The reader is referred to Narasimhan and Kafatos (2016) for details. In brief, it is the *availability of 'which path' information*, not the *act of measurement* itself, that triggers the wave function collapse and determines the specific outcome – whether an interference pattern is

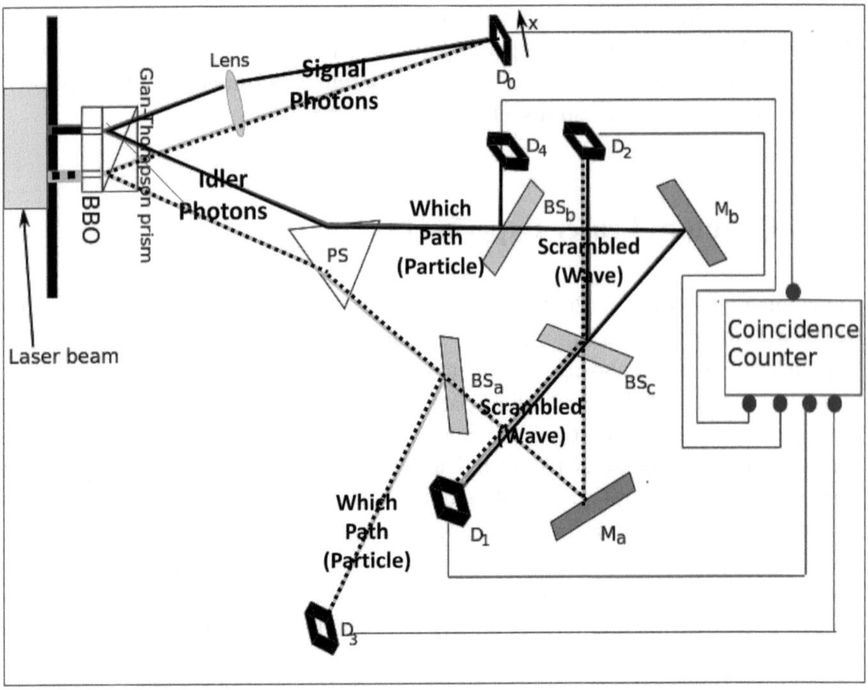

Fig. 18.1 Setup of the delayed choice quantum eraser experiment of Kim et al. Note that Detector D0 is movable. (Adopted from https://en.wikipedia.org/wiki/Delayed_choice_quantum_eraser), while at the Coincidence Counter $T_3 - T_2 = 8$ ns

shown or not. In this Kim et al. (2000) setup, there is no "probing action" or "asking the question" that is normally required in Orthodox QM to trigger the wave function collapse. Similarly (Narasimhan and Kafatos 2016), there is no "Nature responding to the question posed" that is normally required to trigger collapse. Yet, collapse occurs.

In the Kim et al. experimental setup, the initial "act of measurement" of the signal photon happens at T_0. In this case, it always results in no interference pattern, since there is always which-path information, at T_0. While, the measurement of the idler photon happens at T_1 and T_1 is 8 ns "after" T_0. The "initial signal photon" refers to the detection of the signal photon at D_0 at T_0; while the "modulated signal photon" refers to the "modulation" of the signal photon when mapped to the detection at D_x of its idler photon pair at T_1. Therefore, at T_1, when the signal photons at D_0 are matched with the corresponding idler photons, the signal photons corresponding to the idler photons at $D_1$and D_2 always show interference pattern, since "the which" path information is "erased". Similarly, at T_1, the signal photons corresponding to the idler at D_3 and D_4 show no interference patterns. As Narasimhan and Kafatos (2016) pointed out, this happens only when the information available at T_1 is obtained and superimposed on information available at T_0. The observer ***o*** actually observes the end results only at T_2, which can be minutes or hours (practically an infinite time after T_1, as Wheeler pointed out for cosmological situations). Therefore, the observer is not *observing* or *measuring* anything at T_0 or T_1 and is consequently temporally disassociated from the time when the measurement is made or the "question asked of Nature" or "Nature's" response received, as it happens in ordinary delayed choice experiments. What in fact happens at T_1 is that the *information available overrides* the *information available* at T_0 and is then observable by ***o*** at T_2.

In practical situations, and in order to avoid any possible ambiguity concerning the quantum versus classical interpretation, most experimenters (e.g. in Kim et al. 2000) have opted to use non-classical entangled-photon light sources to demonstrate quantum erasers with no possibility of a classical analog.

The delayed choice quantum eraser experiments investigate the following paradox: If a photon manifests itself as though it had come by a single path to the detector, then "common sense" (i.e. classical sense) would require that it must have entered the double-slit device as a particle. Yet, if a photon manifests itself as though it had arrived by two indistinguishable paths, then it must have entered the double-slit device as a wave! In fact, if the experimental apparatus is changed while the photon is in mid-flight, whatever that means in a quantum situation, then the photon should reverse its original "decision" as to whether to behave as a wave or as a particle. The situation illustrated in Fig. 18.1 could apply to cosmological dimensions as Wheeler pointed out: A last-minute "decision" made on earth on how to observe a photon could alter a "decision" made millions or even billions of years ago. It would seem that the past is entangled with the future through the act of measurement, which gives an observer information that the act of measurement, as in delayed choice quantum eraser, made manifest. This is surely not a classical situation.

18.3 Observer and Access to Information in the Quantum Universe

We then have to proceed beyond what the experiments reveal to possible interpretation of the quantum reality. As such, the measurement problem in QM is the question of how (in fact *whether*) the expected wave function collapse occurs. As Weinberg and others (1998) have eloquently pointed out, what is the correspondence between the classical world of experiences and underlying quantum reality? It is of course the inability to directly observe "collapse" that has given rise to a plethora of different interpretations of QM, an unsatisfactory situation.

The question then arises: *Information*, what does the concept really mean? How does it apply in the quantum universe? Where does information reside? These are some of the most important questions that one can pose. We of course know that information is essential for making sense of actual experimental results. We also know that Information plays a crucial role in determining the results of the Measurement (see definitions below).

We also note that, Measurement and *Observation* are *not* the same. In the interpretation of the experimental results discussed here, we provide an initial set of definitions which we will later revise. Leaving aside for now the issue of who the observer is, we have the following:

Observation is availability of information to an "observer".
Measurement is the determination of whether there is which-path information available (particle aspect) or not (wave aspect) to the observer.
Recording is the physical act of documenting the results of the measurement by the observer.

We now allow for the possible refinements to the initial set provided here: The results of the experiment in the Kim et al. setup can be extended to the simple situation illustrated in Fig. 18.2, as follows:

- In the Copenhagen interpretation, the role of the human observer ***o***, as articulated in Sect. 18.1 above, we emphasize here, it is assumed that human observers are essential, and they can be considered to be local observers. In what follows we denote such (human) observers by lower case ***o***. Such a human observer ***o*** sets up the specific quantum apparatus needed to make an observation or 'probing action', in other words "to ask a question of Nature".
- Based on the refinements based on the Kim et al. (2000) setup, we would redefine as follows: The human observer ***o*** is required *only* when: setting up the experiment (shown as the "Grandfather" in Fig. 18.2, which would emphasize from an appreciable time interval in the past); and when reading the recording ("Granddaughter "in Fig. 18.2, which would emphasize some appreciable time interval in the future).

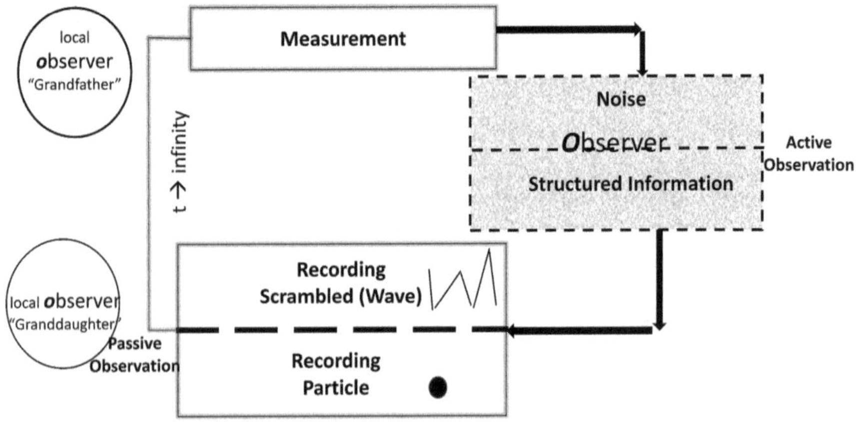

Fig. 18.2 Observation, measurement and recording

- In other words, or for all practical purposes, these two "events" can take place separated by time at values ~ infinity, as Wheeler postulated.
- However, if the human observer ***o*** is *not present* while the experiment is running, i.e., not present when Observation, Measurement and Recording are occurring, then the actual situation means that the presence or not of a human observer *does not change the outcome of the experiment in any way.*
- Therefore, it is clear that the 'Observer' who *influences the outcome* (see definition of *Dynamic, D*, below) is *not* the human observer ***o***.

There is the possibility that *Information* resides in the quantum realm of possibilities, a non-local realm. In future work we will explore the nature of what we will term "Quantum Information Space" (QIS).

Since the Observation and Measurement take place without the need for the human observer to be present in the same space-time coordinates as the one in which the actual experiment is running, it is apparent that the (universal) Observer is non-local, and is, therefore, "outside" space-time, noting however that "outside" is itself a classical space-time concept. We denote this Observer as ***O***.

The question then arises: What are the functions of the Observer ***O***? Why is such an Observer even needed? We have already hypothesized that such an Observer is "outside" of classical space-time information. Such an observer needs to be conscious, in order to distinguish between noise and information, the two complementary aspects revealed in quantum delayed choice eraser experiment. Observer ***O*** is also "dynamic" because it influences the information that is available, *it influences the results of the recording based on whether information is available or not.*

Therefore, ***O*** is:

- Independent, ***I*** (is not part of the experimental system or needing a human observer to be present)
- Conscious, ***C*** (can tell the difference between information and noise)
  Dynamic, ***D*** (has the ability to affect the outcome of recording).

While the ***o*** observer, is a human observer. We emphasize that recording of end results occurs and that the ***O***bserver (in the von Neumann sense) is not the information itself.

A local observer ("Grandfather") at some time in the past, sets up the experiment. This can occur at any time *before* the experiment starts and therefore has no impact on the results of the experiment. Only what happens in the Active Observation box, influences the results of the experiment.

In Fig. 18.2, the experiment starts with the signal photons at T0 and T2 as in the Kim et al. experiment (going through the Observation ➔ Measurement ➔ Recording stages) and continues until their respective idler photons at T1 and T3 (going through the Observation ➔ Measurement ➔ Recording stages), with T1 being 8 ms after T0, and T3 being 8 ms after T2 (see Fig. 18.1). This is the Active Observation phase (or Dynamic phase), where the recorded results are derived/ influenced by the interpretation of whether 'which-path' information is available (recorded as particle) or 'which-path' information is *not* available (recorded as wave). Note again that the recorded results do *not* require the presence of a human observer ***o*** during the time between T0 and T3.

In our view, the Observer is not the grandfather or the granddaughter, since we have just shown that their absence does not change the recorded results.

There is a "black box" independent of the information (shown in Fig. 18.2 in dashed lines and colored green). It is in this 'black box' that the Active Observation takes place and causes the recorded results. It is in this 'black box that the (global) ***O***bserver operates.

Let us examine the characteristics of the Active Observation black-box, in which the Observer operates, that are required to explain the experimental results.

It is obvious that it is Independent **I** of the persons (or the local observers) involved in the experiment. ***O*** is not the human experimenter (grandfather / granddaughter), such an observer is **I, C, D.** The (Global) ***O***bserver is non-local ("outside" space-time), in the Quantum Information Space, QIS, whereas ***o*** is local (in space-time).

In the non-local universe, time does not exist. As such, the scrambling appears to "occur in the future". However, the absence or presence of information is independent of time. This further strengthens the possibility that the Observer is independent of space-time.

Therefore, Measurement is the *Observer observing* if there is which-path information available or not. Recording is the *physical documenting* of the results of the Measurement by the Observer and so it a 'mechanical' operation that does not need any participation. The ***o***bserver does not need to participate in either Measurement or Recording, as the experiment "runs by itself". The only role of the ***o***bserver (Granddaughter) is to read the Recording at a future point of time.

In view of the above, **Measurement** is making the information available to the Observer. **Observation** is the Observer determining if which-path information is available or not, and then dynamically changing the recorded results based on that observation. Finally, **Recording** is the physical act of documenting the results of the dynamic output from the Observer. None of these steps requires the presence of the ***o***bserver.

From the above, it is clear that what determines the end result (Recording) of either the Wave/interference pattern or the Particle pattern, is only the availability of information.

In the "quantum eraser" focus of the experimental design, the design "erases" the which-path information for certain tracks. The question is whether the act of erasure still constitutes "measurement". Based on the design of the experiment, we suggest that it does.

However, after the "measurement" that causes erasure of information, the recorded result is that of a wave. We reiterate that it is, therefore, clear that it is not the act of measurement itself that causes the collapse of the wave function, but the presence of information (specifically, the presence of which-path information in this case which results in the observation of a particle). When the which-path information is erased in a separate act of measurement, it results in the observation of a wave.

It is *not*, as commonly assumed, the presence of a human observer making observations, which determines the end result, i.e. the "collapse of the wave function". It is clear from the results of the experiment that the presence in space-time of the human observer ***o*** (grandfather or granddaughter) is not required and does not influence in any way, the determination of the recorded outcome.

Furthermore, the results if the "delayed quantum eraser" focus of this experiment is that the results recorded at time T1 and T3, change the results "retroactively"' recorded at T0 and T2, which are 8 ms *before* T1 and T3.

These experimental results seem to indicate two seeming paradoxes a) not requiring a human observer ***o*** to change the results of the experiment and b) "retroactivity" (or non-locality). We suggest that these can be explained by postulating an Observer ***O*** that is not in space-time, non-local.

To reiterate: The Universal Observer (non-local) is not in space-time. We denote this Observer as ***O***. Such an Observer is "outside" of information. ***O*** needs to be conscious, in order to distinguish between noise and information. Observer ***O*** is "dynamic" because it influences the information that is available. It influences the recording. Therefore, to summarize, ***O*** is:

- Independent, **I** (is not part of the experimental system or needing a human observer to be present)
- Conscious, **C** (can tell the difference between information and noise)
- Dynamic, **D** (has the ability to affect the outcome of recording).

While ***o*** is a human observer, in space-time.

18.4 Discussion and Conclusions

The work presented here is continuation of our previous research presented in Narasimhan and Kafatos (2016). We may inquire about further empirical support for the ideas presented in their work and here. As pointed out in Kafatos and Narasimhan (2016), we note that several quantum eraser experiments have been carried out. The existence of ***O*** and ***o*** is obtained from the type of delayed quantum eraser experiment, which provides the experimental evidence that fits with, and is consistent with, our hypothesis. We also note if confirmation is sought in the classical domain, then the normal double slit experiment, without delayed quantum erasure, would show what happens in the classical world. In the "outside" space-time domain, there is paradox, since there is no "time" separating cause and effect.

When information is unstructured, it is "wave" (actually it is "noise", or a set of infinite possibilities expressed in the wave function). A set of infinite possibilities expressed in the wave function cannot be located in a specific region in space-time and so needs to be thought of as not existing in space-time, or "outside" of space-time.

In our view, observation is the ability to distinguish between random noise and structure which in turn implies an Observer ***O***. Recognition of structured information implies the ability to recognize information (Kafatos and Nadeau 2000). This would render the means to explore what has been termed Fundamental Awareness (Theise and Kafatos 2016).

What is then the famous "collapse" in Orthodox QM? In our view, it is the manifestation of wave in space-time as particle. Such particles are manifest in space-time (because a particle is "local"); and all probabilities collapse into one actuality. This is why we cannot have a particle (which by definition is local) in *(Non-Local) Quantum Information Space* (QIS).

The relationship space-time and QIS would be of the nature of complementary constructs and as such, there would need to be a "leakage" and interaction between the two (complementary) domains, since the Observation is made in (Non-Local) QIS but the results are manifest in local space-time. ***O*** is tied to **o** (or non-local to local). The dynamics of this interaction will be a subject for future study.

As in our previous work, Narasimhan and Kafatos (2016), the approach we take here is agnostic as to the role of Nature and possible modifications to standard Orthodox quantum mechanics (Stapp 2011). The roles of ***O*** and **o** presented here are novel interpretations, consistent with the advanced eraser experiments of Kim et al. (2000).

In this work, we have expanded on the earlier ideas to introduce an additional key attribute to the Non-Local Observer ***O***. This is concept of **D** as described in the earlier section – the ability of the Observer ***O*** to be Dynamic, ie able to somehow influence the outcome of the results and not be merely a passive observer.

This has two implications: firstly, it implies that ***O***, which is non-local, has the ability to not merely measure the information, but to dynamically and directly influence the results which are recorded in space-time. Secondly, it implies that ***O*** in some way can 'bridge' and work in both non-local and local frames of reference.

We actually believe that von Neumann [20] himself implied the existence of such an Observer as he concluded that the so-called 'Heisenberg Cut' was probably nowhere to be found. 'Nowhere', of course, implies 'not located in local space-time'. In our view that would be consistent with the existence of ***O***.

The Conscious Observer is able to interpret in a time independent way (e.g. independent of T_0, T_1), and as such the Conscious Observer has to be *outside space-time*, i.e., or *non-local*. We also developed the ideas for measurement, recording; as well as passive versus active observation, all as part of the observation process and the access to information. Non-locality is an incontrovertible and significant quantum phenomenon (Chiao et al. 1995; Kafatos and Nadeau 2000) that presents a totally different paradigm than local, classical reality.

H.P. Stapp has advanced our understanding of quantum phenomena and the role of the mind (cf. Stapp 2009). In the Orthodox interpretation of quantum mechanics (Stapp 2009), which extends the Copenhagen Interpretation as achieved by von Neumann (1995), the role of observation played by an observer and the response by Nature are examined and outlined (Stapp 2007). Although we agree with the basic approach of the role of the mind in the Copenhagen Interpretation and its Orthodox refinement by von Neumann, we have shown in Kafatos and Narasimhan (2016) and here that the issues are more complicated when quantum eraser non-locality enters the picture as collapse occurs without an act of observation by a local observer; quantum eraser experiments, imply, although cannot "prove", as "proof" would require that everything takes place in space-time, in our view the availability of information outside of space-time and are consistent with theoretically and experimentally established quantum non-locality (Aspect 1999).

We realize that our interpretation opens up a set of issues that need further exploration. What is the relationship between local observers and the universal Observer? What is the Quantum Information Space? What is information itself? Is it objectively defined?

In future work, we will examine the interaction between Non-local Observer and local observer: **o** – that is the local conscious **observer** that exists *in space-time*, and is an integral part of ***O***. It may indeed be the case that ***O*** and **o** are related, and together, they are the bridge between space-time *manifestation* and outside space-time *potential* (QIS). We will also examine the full implications and how these points outlined here may be in agreement with non-physical views of reality (Theise and Kafatos 2016) and related mathematical models (Kafatos 2015). What we suggest here, does not negate standard Orthodox QM. On the contrary, it extends it and we believe, makes it the dominant view of quantum reality.

References

Aspect, A. (1999). Bell's inequality test: More ideal than ever. *Nature, 398*, 189.

Bartell, L. (1980). Complementarity in the double-slit experiment: On simple realizable systems for observing intermediate particle-wave behavior. *Physical Review D, 21*(6), 1698–1699. https://doi.org/10.1103/PhysRevD.21.1698. Bibcode:1980PhRvD..21.1698B.

Boscá Díaz-Pintado, M. C. (2007). *Updating the wave-particle duality*, 15th UK and European meeting on the foundations of Physics, 29–31 March, 2007. Leeds, UK.

Cassidy, D. (2008). *Quantum mechanics 1925–1927: Triumph of the Copenhagen interpretation* (Werner Heisenberg, Ed.). New York: American Institute of Physics.

Chiao, R.Y., Kwiat, P.G., & A. M. Steinberg, A.M. (1995). Quantum non-locality in two-photon experiments at Berkeley, Quantum and Semiclassical Optics: Journal of the European Optical Society Part B 7 (3): 259–278. https://doi.org/10.1088/1355-5111/7/3/006. arXiv:quant-ph/9501016. Bibcode:1995QuSOp...7..259C

Donati, O., Missiroli, G. F., & Pozzi, G. (1973). An experiment on Electron interference. *American Journal of Physics, 41*, 639–644. https://doi.org/10.1119/1.1987321. Bibcode:1973AmJPh..41..639D.

Feynman, R. P., Leighton, R. B., & Sands, M. (1965). *The Feynman lectures on physics* (Vol. 3, pp. 1.1–1.8). San Francisco: Addison-Wesley. ISBN:0-201-02118-8.

Harrison, D. (2002). *Complementarity and the Copenhagen interpretation of quantum mechanics UPSCALE*. Department of Physics, University of Toronto.

Herzog, T. J., Kwiat, P. G., Weinfurter, H., & Zeilinger, A. (1995). Complementarity and the quantum eraser. *Physical Review Letters, 75*(17), 3034–3037. https://doi.org/10.1103/PhysRevLett.75.3034. Bibcode:1995PhRvL..75.3034H. Retrieved February 13, 2014.

Kafatos, M. C. (2015). Fundamental mathematics of consciousness. *Cosmos and History: The Journal of Natural and Social Philosophy, 11*(2), 175–188. http://www.cosmosandhistory.org/index.php/journal

Kafatos, M., & Nadeau, R. (1990/2000). *The conscious universe: Parts and wholes in physical reality* New York: Springer.

Kafatos, M. C., Narasimhan, A. (2016). Mathematical Frameworks for Consciousness, *Cosmos and History: The Journal of Natural and Social Philosophy, 12*(2), 150–159. http://www.cosmosandhistory.org/index.php/journal/article/view/554/905

Kim, Y. H., Yu, R., Kulik, S. P., Shih, Y. H., & Scully, M. (2000). A delayed "Choice" quantum eraser. Physical Review Letters, *84*, 1–5. https://doi.org/10.1103/PhysRevLett.84.1. arXiv:quant-ph/9903047. Bibcode: 2000PhRvL..84....1K.https://en.wikipedia.org/wiki/Delayed_choice_quantum_eraser

Narasimhan, A., & Kafatos, M. C. (2016). Wave particle duality, the observer and retrocausality, *Quantum Retrocausation III,* Daniel P. Sheehan (Ed.) AIP conference proceedings, 1841:040004-1, 9. 9 pages.

Scully, M. O., & Drühl, K. (1982). Quantum eraser: A proposed photon correlation experiment concerning observation and "delayed choice" in quantum mechanics. *Physical Review A, 25*(4), 2208–2213. https://doi.org/10.1103/PhysRevA.25.2208. Bibcode:1982PhRvA..25.2208S.

Stapp, H. P. (2007). *Mindful universe: Quantum mechanics and the participating observer* (The frontiers collection) (2nd ed.). Berlin/Heidelberg: Springer.

Stapp, H. P. (2009). *Mind, matter, and quantum mechanics*. Berlin/Heidelberg: Springer.

Stapp, H.P. (2011). Retrocausal effects as a consequence of orthodox quantum mechanics refined to accommodate the principle of sufficient reason. In *Quantum retrocausation: Theory and experiment*, 13–14, June 2011, San Diego, CA, AIP conference proceeding 1408, D. Sheehan (Ed.), pp. 31–44. https://doi.org/10.1063/1.3663730.

Theise, N. D., & Kafatos, M. C. (2016). Fundamental awareness: A framework for integrating science, philosophy and metaphysics. *Communicative & Integrative Biology, 9*(3), e1155010. https://doi.org/10.1080/19420889.2016.115501000-00.

von Neumann, J. (1995). *Mathematical foundations of quantum theory*. Princeton: Princeton University Press.

Walborn, S. P., et al. (2002). Double-slit quantum eraser. *Physical Review A, 65*(3), 033818. https://doi.org/10.1103/PhysRevA.65.033818. arXiv:quant-ph/0106078. Bibcode:2002PhRvA..65c3818W.

Weinberg, S., Howard, M., & Roger Louis, W. (Eds.). (1998). *The Oxford history of the twentieth century* (p. 26). Oxford: Oxford University Press. ISBN:0-19-820428-0.

Wheeler, J. A. (1981). *In some strangeness in the proportion* (H. Woolf, Ed.). Reading: Addison-Wesley Publishing.

Zajonc, A. G., Wang, L. J., Zou, X. Y., & Mandel, L. (1991). Quantum eraser. *Nature, 353*, 507–508. https://doi.org/10.1038/353507b0.

Chapter 19
Unifying Decision-Making: A Review on Evolutionary Theories on Rationality and Cognitive Biases

Catarina Moreira

Abstract In this paper, we make a review on the concepts of rationality across several different fields, namely in economics, psychology and evolutionary biology and behavioural ecology. We review how processes like natural selection can help us understand the evolution of cognition and how cognitive biases might be a consequence of this natural selection. In the end we argue that humans are not irrational, but rather rationally bounded and we complement the discussion on how quantum cognitive models can contribute for the modelling and prediction of human paradoxical decisions.

Keywords Rationality · Cognitive bias · Evolutionary biology · Behavioural ecology · Quantum cognition · Decision-making

19.1 Introduction

Rationality is one of the oldest and yet still on going research topics in the scientific community. Specially in the modern world where computational tools are used for predictive analyses of human behaviour and for the development of more sophisticated decision-making models that are able to contribute for a general theory for decision-making. Rationality is a term that has been widely debated across several fields of the literature with different discussions in Economics, Psychology and behavioural Ecology/evolutionary Biology.

In 1944, the Expected Utility theory was axiomatised by the mathematician John von Neumann and the economist Oskar Morgenstern, and became one of the most significant and predominant rational theories of decision-making (von Neumann and Morgenstern 1953). The Expected Utility Hypothesis is characterised by a specific set of axioms that enable the computation of the person's preferences

C. Moreira (✉)
School of Information Systems, Faculty of Science and Technology, Queensland University of Technology, Brisbane, QLD, Australia
e-mail: catarina.p.moreira@tecnico.ulisboa.pt

J. Acacio de Barros, C. Montemayor (eds.), *Quanta and Mind*, Synthese Library 414, https://doi.org/10.1007/978-3-030-21908-6_19

with regard to choices under risk (Friedman and Savage 1952). By risk, we mean an uncertain event that can be measured and quantified. In other words, choices based on *objective probabilities*. Under this theory, human behaviour is assumed to maximise an utility function and by doing so, the person would be acting in a fully rational setting. This means that human psychological processes started to be irrelevant as long as human decision-making obeys to some set of axioms (Glimcher and Fehr 2014).

In 1953, Allais proposed an experiment that showed that human behaviour does not follow these normative rules and violates the axioms of Expected Utility, leading to the well known Allais paradox (Allais 1953). Later, in 1954, the mathematician Leonard Savage proposed an extension of the Expected Utility hypothesis, giving origin to the Subjective Expected Utility (Savage 1954). Instead of dealing with decisions under risk, the Subjective Utility theory deals with *uncertainty*. Uncertainty usually described situations that involve ambiguous/unknown information. Consequently, it is specified by subjective probabilities. In 1961, Daniel Ellsberg proposed an experiment that showed that human behaviour also contradicts and violates the axioms of the Subjective Expected Utility theory, leading to the Ellsberg paradox (Ellsberg 1961). In the end, the Ellsberg and Allais paradoxes show that human behaviour does not follow a normative theory and, consequently, tends to violate the axioms of rational decision theories. In summary, when dealing with preferences under uncertainty, it seems that models based on normative theories of rational choice tend to tell how individuals *must* choose, instead of telling how they *actually* choose (Machina 2009).

The separation of economics from psychology made these two research fields take their own separate paths in terms of human decision-making. In one extreme, economics was built up from a set of strong normative assumptions that assume that to be fully rational means to obey to some set of axioms. Everything that is not chosen according to these normative axioms lead to irrational behaviour.

In this paper, we will put together some major scientific contributions from different fields that could help to unify interdisciplinary knowledge towards a more unified decision model.

19.2 The Different Types of Rationality

The psychologist Gerd Gigerenzer makes a distinction between the degrees of rationality that a theory of decision-making should incorporate (Gigerenzer and Selten 2001). Disciplines like Economics, Cognitive Science, Biology, etc., assume that both humans and animals have unlimited information, unlimited computational power and unlimited time to make a decision. And consequently, as long as these decisions follow the axioms of the expected utility theory, then they are optimal and fully rational. This kind of rationality is called *unbounded rationality*. Figure 19.1 shows the different kinds of rationality that one can find across different disciplines as proposed by Gigerenzer and Selten (2001). Figure 19.1, makes a distinction

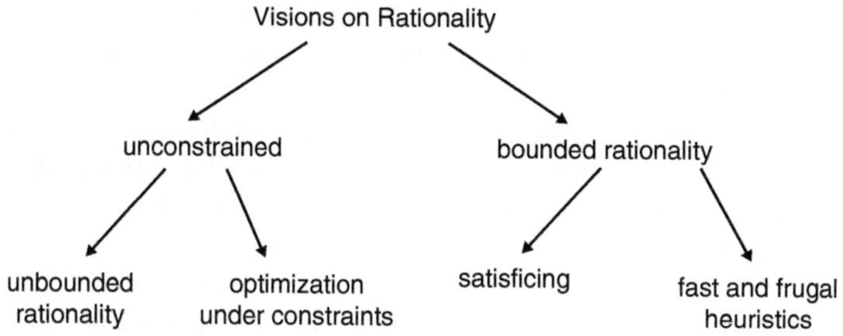

Fig. 19.1 Different types of rationality across different fields as suggested in the work of Gigerenzer and Selten (2001)

between two unconstrained theories of rationality: the unbounded rationality and the optimisation under constraints. The difference is that the later has a stopping rule, which measures and stops at the point where the cost for further search exceeds the benefits of the decision.

Although unbounded rationality is widely used across several fields (specially in Economics), it is not a concept that can be applied in real world decision scenarios. Both humans and animals have limited resources: they need to make inferences under uncertain aspects of the world with limited knowledge, limited computational power and limited time. For instance, an animal that is exposed in the wild looking for food needs to make the decision if it will either remain looking for food being completely exposed to his predators or if he should hide. If he has unlimited time to make this decision, this could evolve the computation of several alternatives in order to choose the one that grants the animal the highest utility. However, if the animal looses too much time in this reasoning process, he could get attacked by its predators. So, there are constraints on resources when making decisions under uncertainty. One can even argue that both humans and animals are fully rational under these constraints: we try to find the best option with the limited time, computational power and information that we have. This is usually called *bounded rationality* and usually incorporates heuristics (or rules of thumb) which are gathered by experience and memory in order to turn the decision process faster. Bounded rationality was a term introduced by Simon (1955) where he suggested that decision-makers are seen as satisficers, who seek a solution that is satisfiable under constraints in time, information and computation, rather than the optimal one.

19.3 Biological System: Does Natural Selection Produce Rational Behaviour?

In evolutionary Biology, the *fitness* of an individual is a measure how good an individual (or a species) is at leaving offsprings in the next generation. This is a very important concept, since natural selection consists on the heritable variations that a trait can suffer and consequently influence the fitness level of an individual (Stevens 2008). The premise is that, the more surviving descendants an individual produces, the higher is the fitness level.

Since natural selection is the process by which biological evolution occurs, leading to a species with more fitness, many scholars have posed the question whether natural selection has any role in rational behaviour (Stevens 2008; Houston et al. 2007; Santos and Rosati 2015).

Like humans, animals also need to make decisions that grants them higher profits. In other words, they need to take into account temporal delays and uncertainty as well as potential payoffs in pursuing different actions (Santos and Rosati 2015). In this sense, human decisions under economic scenarios can have some analogies with the type of decisions that animals are faced when they are foraging for food or seeking mates: in both scenarios it is assumed that they prefer choices that grant them higher profits/food/fitness levels. The question that arises in this context is if, under an evolutionary point of view, natural selection leads to choices that are in accordance with the expected utility hypothesis, then how did human biases emerge?

One of the cognitive biases pointed in Tversky and Kahneman (1986) work is the *framing effect*. In framing effects, people react differently to some choice depending on how that choice is presented. For instance, people tend to avoid risk when a positive frame is presented but seek risks when a negative frame is presented (Kahneman et al. 1982). Research shows that these framing effects are not singular to humans, but they also occur in primates. For instance, Chen et al. (2006) made an experiment where monkeys could trade tokens with human experimenters in exchange for food. In the end, the amount of food that each monkey received form the human experimenters was the same. The difference was that one experimenter showed the monkey one piece of apple and then added an extra one (the gains experimenter), while the other showed two pieces of apple to the monkey, but removed one (the losses experimenter). Although the monkeys received the same payoff, they preferred the gains experimenter over the losses one, indicating that monkeys also fall under the framing effect like humans (Santos and Rosati 2015).

Another study with monkeys conducted by Lakshminarayanan et al. (2011) showed that monkeys also revealed a reflection effect. The reflection effect explains that we have opposite 'risk preferences' for uncertain choices, depending on whether the outcomes is a possible gain or a loss. In the study of Lakshminarayanan et al. (2011), monkeys tended to seek out more risk when dealing with losses when compared to gains.

These works suggest that cognitive biases are not singular to humans, but they occur in non-human beings as well. These findings could imply some ancestral roots in our cognitive system. Returning to the question of whether natural selection plays a role in a fully rational cognitive system, studies suggest that it is not the case (Stevens 2008). Natural selection does play a role in optimising the fitness of an individual, like it is stated in the work of Santos and Rosati (2015). If the environment is constant, then the process of natural selection of a species reaches its optimal fitness with time. It seems that context does play an important role in natural selection.

Following the discussion on Santos and Rosati (2015), in evolutionary biology it is agreed that it may be rational for an individual to make biased and inconsistent preferences, if these preferences point towards the maximisation of the fitness (Kacelnik 2006). In other words, individuals my be acting rational by falling into cognitive biases that may lead them to the survival of the species. This can be connected with the notion of bounded rationality that it was presented in the previous section. These cognitive biases produce rules of thumb for individuals that, although they are not optimal, they lead to the best possible decision given the limited constraints on time, computational power and information. In other words, they are able to reach a decision that grants them a high return (not the most optimal one), using less computational resources through the usage of heuristics. Note that, the notion of heuristic consists precisely in a *shortcut* that usually leads to the desired outcome, but sometimes and get lead to wrong decisions and outcomes (Shah and Oppenheimer 2008).

Experiments conducted on birds also show the importance of context in behavioural ecology. Studies have shown that both birds and insects deviate significantly from their choices when the context varies (Kacelnik and Marsh 2002). Following the arguments in Santos and Rosati (2015), in behavioural ecology, the energetic increase of an individuals fitness level is non-linear. This means that a single unit of food has a significant impact on an individual that has a low level of energy (an individual that is hungry), but it would have a lower impact (or even a negative impact) on an individual that is already in a high energy state (not hungry). So, the risk preferences that both humans and animals choose in different contexts might be optimising the fitness measure under a biological point of view depending on the context where they are.

Summarising, natural selection alone does not lead to optimal rational and cognitive behaviour, but it always leads to the optimisation of the fitness measure. If the context (or environment) where the individual is contained is constant, then natural selection can indeed lead to fully rational and optimal choices. However, these are exceptional scenarios. Throughout time, environments keep changing and individuals are required to adapt in order to optimise their fitness measure and to survive. Under a constant change of context and environment, individuals might be more susceptible to fall under cognitive bias and apply rules of thumb and heuristics that can help them get the decision outcome taking into considerations the limitations in knowledge regarding the new environmental context, limitations in computational power and limitations in time. Ultimately, individuals are not being irrational, but rather rational given all these constraints. In economics, cognition and other disciplines, this behaviour is seen as purely irrational, since it is violating the axioms of expected utility theory.

The question that now arises is whether there is a mathematical theory that is suitable to mode these cognitive biases and violations to the axioms of expected utility in a more general and elegant way that could be used across different fields that analyse cognition and decision-making through different perspectives. We suggest that the answer to this question is positive and that the answer can be pointed towards the mathematical formalisms of quantum mechanics.

19.4 Quantum Cognition

Motivated by cognitive biases identified by Tversky and Kahnman, researchers started to look for alternative mathematical representations in order to accommodate these violations (Tversky and Kahneman 1974, 1986; Kahneman and Tversky 1972; Tversky and Shafir 1992). Although in the 40s, Niels Bohr had defended and was convinced that the general notions of quantum mechanics could be applied in fields outside of physics (Murdoch 1989), it was not until the 1990s that researchers started to actually apply the formalisms of quantum mechanics to problems concerned with social sciences. It was the pioneering work of Aerts and Aerts (1994) that gave rise to the field *Quantum Cognition*. In their work, Aerts and Aerts (1994) designed a quantum machine that was able to represent the evolution from a quantum structure to a classical one, depending on the degree of knowledge regarding the decision scenario. The authors also made several experiments to test the variation of probabilities when posing *yes/no* questions. According to the authors, the experiment suggested that when participants did not have a predefined answer regarding a question, then the answer would be updated at the moment the question was asked, indicating that the answer is highly contextual. In other words, the answer is formed by the interaction between the participant and the person asking the question. This deviates from classical statistics, because the answer is not dependent on the participant's beliefs. A further discussion about this study can be found in the works of Aerts (1995, 1996, 1998), Gabora and Aerts (2002), and Aerts et al. (2011).

Quantum cognition has emerged as a research field that aims to build cognitive models using the mathematical principles of quantum mechanics. Given that classical probability theory is very rigid in the sense that it poses many constraints and assumptions (single trajectory principle, obeys set theory, etc.), it becomes too limited (or even impossible) to provide simple models that can capture human judgments and decisions since people are constantly violating the laws of logic and probability theory (Busemeyer 2015; Busemeyer and Wang 2014; Aerts 2014).

Following the lines of thought of Sloman (2014), people have to deal with missing/unknown information. This lack of information can be translated into the feelings of ambiguity, uncertainty, vagueness, risk, ignorance, etc. (Zadeh 2006), and each of them may require different mathematical approaches to build adequate cognitive/decision problems. Quantum probability theory can be seen as an alternative mathematical approach to model such cognitive phenomena.

The heart of quantum cognition is to use concepts of quantum mechanics such as superposition and quantum interference effects in order to accommodate the paradoxical findings found in the literature. For instance, in quantum information processing, information is modelled via wave functions and therefore they cannot be in definite states. Instead, they are in an indefinite quantum state called the *superposition* state. That is, all beliefs are occurring on the human mind at the same time, instead of the classical approach which considers that each belief occurs in each time frame. According to cognitive scientists, this effect is responsible for making people experience uncertainties, ambiguities or even confusion before making a decision. At each moment, one belief can be more favoured than another, but all beliefs are available at the same time. In this sense, quantum theory enables the modelling of the cognitive system as if it was a wave moving across time over a state space until a final decision is made. From this superposition state, uncertainty can produce different waves coming from opposite directions that can crash into each other, causing an interference distribution. This phenomena called *quantum interference effects* is the heart of quantum cognition and it can never be obtained in a classical setting. When the final decision is made, then there is no more uncertainty. The wave collapses into a definite state. Thus, quantum information processing deals with both definite and indefinite states (Busemeyer and Bruza 2012). Figure 19.2 shows an example of quantum superposition and quantum interference effects.

Some researchers argue that quantum-like models do not offer many underlying aspects of human cognition (like perception, reasoning, etc.). They are merely mathematical models used to fit data and for this reason they are able to accommodate many paradoxical findings (Lee and Vanpaemel 2013). Indeed quantum-like models provide a more general probability theory that use quantum interference effects to model decision scenarios, however they are also consistent with other psychological phenomena (for instance, order effects) (Sloman 2014). In the book of Busemeyer and Bruza (2012), for instance, the feeling of uncertainty or ambiguity can be associated to quantum superpositions, in which assumes that all beliefs of a person occur simultaneously, instead of the classical approach which considers that each belief occurs in each time frame.

The non-commutative nature of quantum probability theory enables the exploration of methods capable of explaining violations in order of effects. Order of effects is a fallacy that consists in querying a person in one order and then posing the same questions in reverse order. Through classical probability theory, it would be expected that a person would give the same answers independently of the order of the questions. However, empirical findings show that this is not the case and

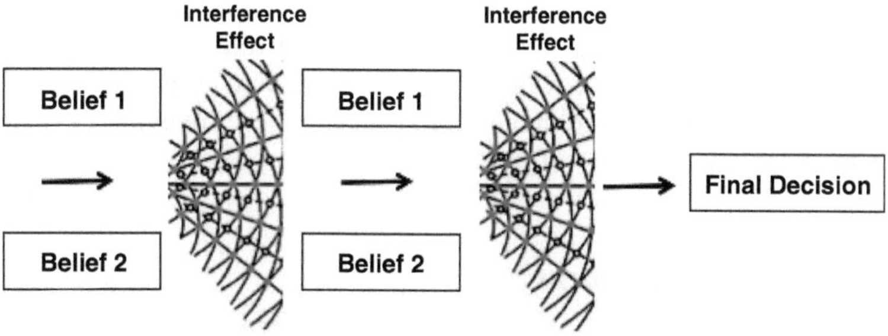

Fig. 19.2 Example of quantum superposition and quantum interference effects. Human beliefs are seen as indefinite states that are occurring in the human mind at the same time. If we model these beliefs as waves that propagate, then they can interfere with each other through quantum interference effects leading to different decision outcomes

that people are influence by the context of the previous questions. Moreover, the existence of quantum interference effects also enables the exploration of models that accommodate other typed of paradoxical findings such as disjunction and conjunction errors (Tversky and Shafir 1992; Tversky and Kahneman 1983). In summary, quantum probability theory is a general framework, which can naturally explain various decision-making paradoxes without falling into the restrictions of classical probability theory.

There are many quantum-like models proposed in the literature (Aerts et al. 2013a,b; Pothos and Busemeyer 2009; Busemeyer et al. 2006b; Khrennikov 2010; Yukalov and Sornette 2011). For the purposes of this paper, we will present a model that is based on modularity. Like it was presented, both humans and non-humans have limited computation and information processing capabilities. In order to reason about something, one needs to combine small pieces of information in order to make a decision about it (Griffiths et al. 2008). One model that can represent this modularity is the Quantum-Like Bayesian Network originally proposed by Moreira and Wichert (2014, 2016).

19.5 Modularity and Quantum-Like Bayesian Networks for Decision-Making

Bayesian Networks are one of the most powerful structures known by the Computer Science community for deriving probabilistic inferences (for instance, in medical diagnosis, spam filtering, image segmentation, etc.) (Koller and Friedman 2009). They provide a link between probability theory and graph theory. And a fundamental property of graph theory is its modularity: one can build a complex system by combining smaller and simpler parts. It is easier for a person to combine pieces of evidence and to reason about them, instead of calculating all possible events and their respective beliefs (Griffiths et al. 2008). In the same way, Bayesian Networks represent the decision problem in small modules that can be combined to perform inferences. Only the probabilities which are actually needed to perform the inferences are computed.

This process can resemble human cognition (Griffiths et al. 2008). While reasoning, humans cannot process all possible information, because of their limited capacity. Consequently, they combine several smaller pieces of evidence in order to reach a final decision.

19.5.1 Bayesian Networks

A classical Bayesian Network can be defined by a directed acyclic graph structure in which each node represents a different random variable from a specific domain and each edge represents a direct influence from the source node to the target node.

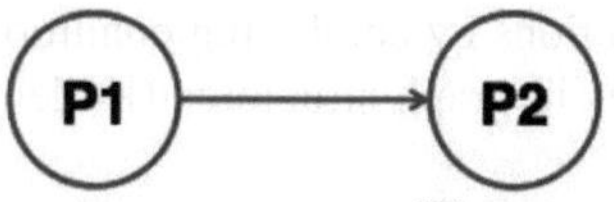

Pr(P1 = defect) = 0.5
Pr(P1 = cooperate) = 0.5

P1	Pr(P2 = defect)	Pr(P2 = cooperate)
defect	0.87	0.13
cooperate	0.74	0.26

Fig. 19.3 Example of classical Bayesian network with two random variables (nodes)

The graph represents independence relationships between variables and each node is associated with a conditional probability table which specifies a distribution over the values of a node given each possible joint assignment of values of its parents. This idea of a node, depending directly from its parent nodes, is the core of Bayesian Networks. Once the values of the parents are known, no information relating directly or indirectly to its parents or other ancestors can influence the beliefs about it (Koller and Friedman 2009). Figure 19.3, shows the representation of a Bayesian Network.

Bayesian Networks are based on probabilities. Probability theory is a formal framework that is capable of representing multiple outcomes and their likelihoods under uncertainty. Uncertainty is a consequence of various factors: limitations in our ability to observe the world, limitations in our ability to model it and possibly even because of innate nondeterminism (Koller and Friedman 2009). If we take into account the basic structures behind probabilistic systems, we will find that in one extreme there is the full joint probability distribution and in another extreme there is the Naïve Bayes model.

The full joint probability distribution corresponds to the entire knowledge of a system. That is, it represents the probabilities of all possible atomic events in a domain. However, this explicit representation of probability distributions is a highly demanding process and most of the times such information is not possible to obtain as a full. Even in its simplest case, when N variables of a distribution contain binary values, one would need to compute 2^N entries. Under a computational point of view, the manipulation of this distribution is a heavy and expensive process and the probability distribution tables become too large to be stored in memory.

In the other extreme, there is the Naïve Bayes model. Reasoning about any realistic domain always requires that some simplifications are made. The very act of preparing knowledge to support reasoning requires that we leave many facts unknown, unsaid or crudely summarised. The Naïve Bayes model assumes that all random variables are independent of each other given a class. This strong independence assumption greatly reduces the computational costs when compared to full joint probability distribution. However, the independence assumption rarely holds when it comes to modelling real world events, leading to more inaccurate models.

Bayesian Networks can be seen as a framework that falls between the full joint probability distribution and the Naïve Bayes Model. It is a compact representation of high dimensional probability distributions by combining conditional parameterisations with conditional independences in graph structures (Koller and Friedman 2009).

A Bayesian Network can be understood as the representation of a full joint probability distribution through conditional independence statements. This way, a Bayesian Network can be used to answer any query about the domain by combining (adding) all relevant entries from the joint probability.

19.5.2 Quantum-Like Bayesian Networks

In the work of Moreira and Wichert (2014, 2016), the authors suggest to define the Quantum-Like Bayesian Network by replacing real probability numbers by quantum probability amplitudes, which are complex numbers.

A complex number is a number that can be expressed in the form $z = a + ib$, where a and b are real numbers and i corresponds to the imaginary part, such that $i^2 = -1$. Alternatively, a complex number can be described in the form $z = |r|\, e^{i\theta}$, where $|r| = \sqrt{a^2 + b^2}$. The $e^{i\theta}$ term is defined as the phase of the amplitude and corresponds to the angle between the point expressed by (a, b) and the origin of the plane. These amplitudes are related to classical probability by taking the squared magnitude of these amplitudes through Born's rule.

The general idea of a Quantum-Like Bayesian network is that when performing probabilistic inference, the probability amplitude of each assignment of the network is propagated and influences the probabilities of the remaining nodes. In other words, *every* assignment of *every* node of the network is propagated until the node representing the query variable is reached. Note that, by taking multiple assignments and paths at the same time, these trails influence each other producing interference effects. If a node (or a random variable) is not observed, then it remains in a superposition state and it can create quantum interference effects that can be used to accommodate the several paradoxical findings in the literature. Figure 19.4 shows the underlying idea of the quantum-like Bayesian Network.

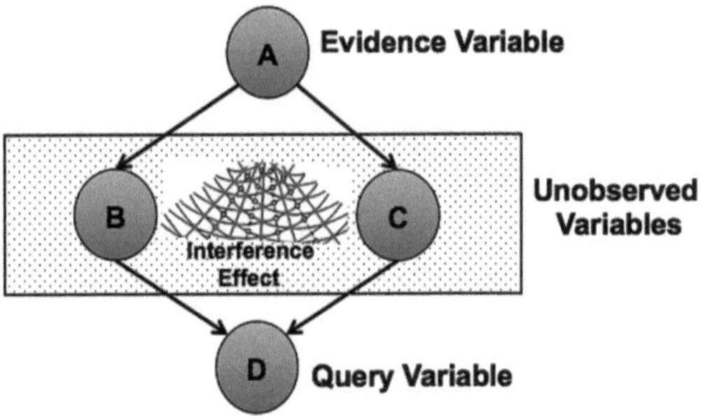

Fig. 19.4 Example of a quantum-like Bayesian network producing quantum interference effects when random variables B and C are not observed

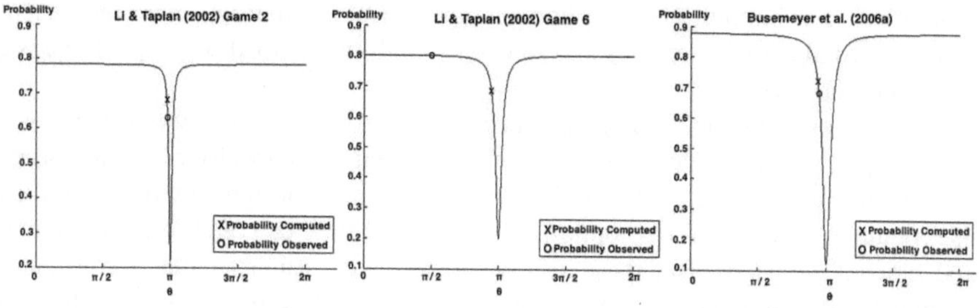

Fig. 19.5 Example of all possible quantum interference effects under the Prisoner's Dilemma experiments made by Li and Taplin (2002) and Busemeyer et al. (2006a). In the figures, it is shown the estimation of the quantum interference terms (*Probability Computed*) through an heuristic function proposed in Moreira and Wichert (2016) and the probability outcome of the participants observed in the experiments (*Probability Observed*)

Recent research shows that quantum interference effects can serve as a fitting function that enables an extra parametric fitting layer when compared with the classical model. Very generally, a probabilistic inference on a quantum-like Bayesian network with two nodes A and B, expressed as probability amplitudes as ψ_A and $\psi_B =$ respectively, is given by

$$Pr(B) = |\psi_A|^2 + |\psi_B|^2 + 2\psi_A\psi_B \cos(\theta_A - \theta_B),$$

where θ_A and θ_B are the quantum interference parameters. Figure 19.5 shows all possible quantum interference effects that can emerge from the Prisoner's Dilemma experiment proposed by Li and Taplin (2002) and Busemeyer et al. (2006a) in order to identify disjunction errors. Note that a disjunction error corresponds to the judgment of two events to be least as likely as either of those events (Carlson and Yates 1989). This way, by choosing the right interference term, one can explain the human cognitive bias and even predict human *rationally bounded* decisions.

19.6 Conclusion

In this work, we made a review of how different fields of the literature deal with the concept of rationality.

Since the axiomatization of the expected utility theory, it has been assumed that humans make optimal decisions, always choosing the preference or the decision that leads to higher gains or higher profits. Every decision that does not maximise this utility is considered irrational and not optimal.

Studies from the last decades show that humans constantly violate the axioms of expected utility theory and now, evolutionary biology and ecological studies also demonstrate that humans are not alone is this subject: both humans and non-humans systematically deviate from what rational choice predicts. If the cognitive bias is shared between humans and non-humans, then it means that we share an ancient and primordial set of heuristics and rules of thumb that helped us in the surviving process of the species.

From the evolutionary Biology perspective, natural selection plays a role in finding the optimal fitness measure. By fitness, we mean the capacity of an individual reproducing and surviving. If an individual remains in a constant environment, then the cognitive abilities will be optimised and indeed they can lead to fully optimal and rational decisions. But these scenarios are scarce. Environment is constantly changing and individuals need to adapt to this change. Given that individuals have limited processing power and limited time, they need to take actions with the limited information that they have. They can use cognitive biases to speed up these decisions. Although they are not fully optimal, they are the best that they can take given the constraints of time, knowledge and computational power. So, under an evolutionary point of view, individuals are rationally bounded, while in disciplines such as economics or cognitive science, these types of decisions are seen irrational.

We ended this paper by introducing quantum mechanics as a general mathematical approach that could take into account quantum superpositions and quantum interference effects to model, explain and predict cognitive biases. We presented a network based model that has been widely studied in the literature that combines small pieces of information in order to perform more complex inferences.

To conclude, there are many disciplines that have large contributions over the topic of rationality and how cognitive biases can be explained through different perspectives. If one wants to model a unified theory of decision-making, one needs to start having a holistic view of the subject in order to produce more robust and more predictive models for decision-making.

References

Aerts, D. (1995). Quantum structures: An attempt to explain the origin of their appearance in nature. *International Journal of Theoretical Physics, 34*, 1–22.

Aerts, S. (1996). Conditional probabilities with a quantal and kolmogorovian limit. *International Journal of Theoretical Physics, 35*, 2245.

Aerts, S. (1998). Interactive probability models: Inverse problems on the sphere. *International Journal of Theoretical Physics, 37*, 305–309.

Aerts, D. (2014). Quantum theory and human perception of the macro-world. *Frontiers in Psychology, 5*, 1–19.

Aerts, D., & Aerts, S. (1994). Applications of quantum statistics in psychological studies of decision processes. *Journal of Foundations of Science, 1*, 85–97.

Aerts, D., Broekaert, J., & Gabora, L. (2011). A case for applying an abstracted quantum formalism to cognition. *New Ideas in Psychology, 29*, 136–146.

Aerts, D., Broekaert, J., Gabora, L., & Sozzo, S. (2013a). Quantum structure and human thought. *Journal of Brain and Behavioral Sciences, 36*, 274–276.

Aerts, D., Gabora, L., & Sozzo, S. (2013b). Concepts and their dynamics: A quantum-theoretic modelling of human thought. *Topics in Cognitive Sciences, 5*, 737–772.

Allais, M. (1953). Le comportement de l'homme rationel devant le risque: Critique des postulats et axiomes de l'École americaine. *Econometrica, 21*, 503–546.

Busemeyer, J. (2015). Cognitive science contributions to decision science. *Cognition, 135*, 43–46.

Busemeyer, J., & Bruza, P. (2012). *Quantum model of cognition and decision* (Cambridge: Cambridge University Press).

Busemeyer, J., & Wang, Z. (2014). Quantum cognition: Key issues and discussion. *Topics in Cognitive Science, 6*, 43–46.

Busemeyer, J., Matthew, M., & Wang, Z. (2006a). A quantum information processing explanation of disjunction effects. In *Proceedings of the 28th Annual COnference of the Cognitive Science Society*.

Busemeyer, J., Wang, Z., & Townsend, J. (2006b). Quantum dynamics of human decision-making. *Journal of Mathematical Psychology, 50*, 220–241.

Carlson, B., & Yates, J. (1989). Disjunction errors in qualitative likelihood judgment. *Organizational Behavior and Human Decision Processes, 44*, 368–379.

Chen, M. K., Lakshminarayanan, V. R., & Santos, L. (2006). How basic are behavioural biases? evidence from capuchin monkey trading behaviour. *Journal of Political Economy, 114*, 517–532.

Ellsberg, D. (1961). Risk, ambiguity and the savage axioms. *Quaterly Economics, 75*, 643–669.

Friedman, M., & Savage, L. (1952). The expected-utility hypothesis and the measurability of utility. *Journal of Political Economy, 50*, 463–474.

Gabora, L., & Aerts, D. (2002). Contextualising concepts using a mathematical generalisation of the quantum formalism. *Experimental & Theoretical Artificial Intelligence, 14*, 327–358.

Gigerenzer, G., & Selten, R. (2001). *Bounded rationality: The adaptive toolbox*. Dahlem workshop reports.

Glimcher, P., & Fehr, E. (Eds.). (2014). *Neuroeconomics: Decision making and the brain*. Amsterdam: Academic/Elsvier.

Griffiths, T., Kemp, C., & Tenenbaum, J. (2008). Bayesian models of inductive learning. In *Proceedings of the Annual Conference of the Cognitive Science Society*.

Houston, A., McNamara, J., & Steer, M. (2007). Do we expect natural selection to produce rational behaviour? *Philosophical Transactions of the Royal Society B, 362*, 1531–1543.

Kacelnik, A. (2006). Meanings of rationality. In S. Hurley & M. Nudds (Eds.), *Rational animals?*. Oxford: Oxford University Press.

Kacelnik, A., & Marsh, B. (2002). Cost can increase preference in starlings. *Animal Behaviour, 63*, 245–250.

Kahneman, D., & Tversky, A. (1972). Subjective probability – a judgment of representativeness. *Cognitive Psychology, 3*, 430–454.

Kahneman, D., Slovic, P., & Tversky, A. (1982). *Judgment under uncertainty: Heuristics and biases*. Cambridge: Cambridge University Press.

Khrennikov, A. (2010). *Contextual approach to quantum formalism*. Springer: Netherlands.

Koller, D., & Friedman, N. (2009). *Probabilistic graphical models: Principles and techniques*. Cambridge/London: The MIT Press.

Lakshminarayanan, V. R., Chen, M. K., & Santos, L. (2011). The evolution of decision-making under risk: Framing effects in monkey risk preferences. *Journal of Experimental Social Psychology, 42*, 689–693.

Lee, M. D., & Vanpaemel, W. (2013). Quantum models of cognition as orwellian newspeak. *Behavioral and Brain Sciences, 36*, 295–296.

Li, S., & Taplin, J. (2002). Examining whether there is a disjunction effect in prisoner's dilemma game. *Chinese Journal of Psychology, 44*, 25–46.

Machina, M. (2009). Risk, ambiguity, and the rank-dependence axioms. *Journal of Economic Review, 99*, 385–392.

Moreira, C., & Wichert, A. (2014). Interference effects in quantum belief networks. *Applied Soft Computing, 25*, 64–85.

Moreira, C., & Wichert, A. (2016). Quantum-like bayesian networks for modelling decision-making. *Frontiers in Psychology, 7*, 11.

Murdoch, D. (1989). *Niels Bohr's philosophy ofphysics*. Cambridge: Cambridge University Press.

Pothos, E., & Busemeyer, J. (2009). A quantum probability explanation for violations of rational decision theory. *Proceedings of the Royal Society B, 276*, 2171–2178.

Santos, L. R., & Rosati, A. G. (2015). The evolutionary roots of human decision-making. *Annual Review in Psychology, 3*, 321–347.

Savage, L. (1954). *The foundations of statistics*. New York: Wiley.

Shah, A., & Oppenheimer, D. (2008). Heuristics made easy: An effort-reduction framework. *Psychological Bulletin, 134*, 207–222.

Simon, H. (1955). A behavioral model of rational choice. *The Quarterly Journal of Economics, 69*, 99–118.

Sloman, S. (2014). Comments on quantum probability theory. *Topics in Cognitive Science, 6*, 47–52.

Stevens, J. (2008). The evolutionary biology of decision-making. In C. Engel & W. Singer (Eds.), *Better than conscious? Decision-making, the human mind and implications for institutions*. Cambridge: MIT Press.

Tversky, A., & Kahneman, D. (1974). Judgment under uncertainty: Heuristics and biases. *Science, 185*, 1124–1131.

Tversky, A., & Kahneman, D. (1983). Extension versus intuitive reasoning: The conjunction fallacy in probability judgment. *Psychological Review, 90*, 293–315.

Tversky, A., & Kahneman, D. (1986). Rational choice and the framing of decisions. *The Journal of Business, 59*, 251–278.

Tversky, A., & Shafir, E. (1992). The disjunction effect in choice under uncertainty. *Psychological Science, 3*, 305–309.

von Neumann, J., & Morgenstern, O. (1953). *Theory of games and economic behavior*. Princeton: Princeton University Press.

Yukalov, V., & Sornette, D. (2011). Decision theory with prospect interference and entanglement. *Theory and Decision, 70*, 283–328.

Zadeh, L. (2006). Generalised theory of uncertainty (gtu) – principal concepts and ideas. *Computational Statistics and Data Analysis, 51*, 15–46.

Chapter 20
"Time Is Out of Joint:" Consciousness, Temporality, and Probability in Quantum Theory

Arkady Plotnitsky

Abstract While the juncture of reality, causality, and probability is a familiar feature of foundational discussions concerning quantum theory, this article considers the role of consciousness and temporality within this juncture, by adopting a nonrealist or, in terms of this article, "reality-without-realism" (RWR) interpretation of it and of quantum phenomena themselves. This interpretation follows Bohr's ultimate interpretation, but possibly takes a more radical epistemological position (the strong RWR view), according to which quantum objects and behavior are not only beyond representation (the weak RWR view), but also beyond conception. The article argues that, at least in this type of interpretation: (a) unlike in classical physics and relativity, the role of consciousness is *irreducible* in quantum experiments insofar as our conscious decisions *determine* the course of future events, rather than merely allow us to *follow* what would happen regardless of such decisions; and (b) that, as a consequence, the workings of temporality require a radical reconsideration in quantum theory, even as against but consistently with relativity and the character of temporality and causality there.

20.1 Introduction

Introduced in 1925, quantum mechanics (QM) and, quickly in its wake, quantum electrodynamics (QED) and quantum field theory (QFT) posed new questions concerning our understanding of the concepts of reality, causality, and probability (referring, respectively, to what exist; how what exists comes about; and how likely something can come into existence) and their relationships, as against classical physics or relativity, or preceding, "old," quantum theory, and led to interpretations of quantum mechanics and quantum phenomena themselves that changed this understanding. Most of these interpretations, beginning with that of N. Bohr,

---

A. Plotnitsky (✉)
Theory and Cultural Studies Program, Purdue University, West Lafayette, IN, USA
e-mail: plotnits@purdue.edu

J. Acacio de Barros, C. Montemayor (eds.), *Quanta and Mind*, Synthese Library 414, https://doi.org/10.1007/978-3-030-21908-6_20

are associated with "the Copenhagen spirit of quantum theory" [*Kopenhagener Geist der Quantenheorie*], as Heisenberg called it (Heisenberg 1930, p. iv). This designation, abbreviated here to "the spirit of Copenhagen," is preferable to the common "Copenhagen interpretation," because there is no single such interpretation even in the case of Bohr, who changed his view a few times (Plotnitsky 2012). In this article, I shall only be concerned with the ultimate version of his interpretation, developed by the late 1930s, following his, by then a decade long, debate with Einstein and responding to "the necessity of a final renunciation of the classical ideal of causality and a radical revision of our attitude towards the problem of physical reality," brought about by quantum phenomena (Bohr 1935, p. 697). In view of this renunciation and this revision: "[T]he recourse to probability laws in quantum physics is essentially different from the familiar application of statistical considerations as practical means of accounting for the properties of mechanical systems of great structural complexity. In quantum physics, we are presented not with intricacies of this kind, but with the inability of the classical frame of concepts to comprise the peculiar feature[s] of the elementary processes," the processes defining the ultimate constitution of nature and responsible for quantum phenomena, defined by the fact that in properly considering then, Planck's constant, h, must be taken into account (Bohr 1987, v. 2, p. 34).

The behavior of all systems considered in classical physics and relativity, or the character of their existence or reality, is assumed to be causal, or in terms adopted here (because there are other concepts of causality), "classically causal:" the state of the system considered is *ontologically* determined at all future moments of time once it is determined at a given moment of time. If the system is sufficiently simple mechanically, it can be handled, *epistemologically*, "deterministically:" our predictions concerning its future behavior could be, ideally, exact. When we deal with "the [classical] mechanical systems of great structural complexity," as in classical statistical physics, while they are still assumed to be classically causal, our predictions concerning them can no longer be deterministic but only probabilistic. In quantum physics, in Bohr's or related interpretations, it is impossible not only to assume the behavior of quantum systems, even the simplest possible ones (such as "elementary particles"), to be causal but also to represent or perhaps even conceive of the ultimate nature or reality of this behavior, which is responsible for the appearance of quantum phenomena. While these interpretations assume the reality of quantum objects they preclude realism in considering this reality, thus compelling me to defining this reality as "reality without realism" (RWR), a concept in accord with the spirit of Copenhagen and specifically Bohr's ultimate interpretation (Plotnitsky 2016; Plotnitsky and Khrennikov 2015). These interpretations (hereafter the RWR-type interpretations) do allow realism at the level of quantum phenomena, described by means of classical physics, which cannot, however, predict these phenomena. In quantum physics, there are no systems, no matter how simple, the behavior of which could be predicted exactly.

The juncture of reality, causality, and probability is a familiar feature of foundational discussions concerning quantum theory. The main contribution of this article is to consider of the role of consciousness and temporality in quantum theory, adopting an RWR-type interpretation of it and of quantum phenomena themselves. This interpretation follows Bohr's ultimate interpretation, but possibly takes a more radical epistemological position (the strong RWR view), according to which quantum objects and behavior are not only beyond representation (the weak RWR view), but also beyond conception. It is not clear whether Bohr have ever held the strong RWR view, and in any event, he never expressly said so. The article will argue that, at least in this type of interpretation: (a) unlike in classical physics and relativity, the role of consciousness is *irreducible* in quantum experiments insofar as our conscious decisions *determine* the course of future events, rather merely allow us to *follow* what would happen regardless of such decisions; and (b) that, in part as a consequence, the workings of temporality, require a radical reconsideration in quantum theory, even as against but consistently with relativity and the character of temporality and causality there. In quantum physics, "time is out of joint," in Hamlet's famous statement ("Time is out of joint. O cursed spite/That ever I was born to set it right" [*Hamlet*, I.v.190–92]). Unlike, however, in Shakespeare's play, where this statement refers to a single special moment of the narrative, in quantum physics, this statement literally applies every time the concept of time applies. To "set it right" is, while possible, means something very different from restoring the spacetime continuum to the ultimate constitution of reality, because, in the RWR-type interpretations, quantum events cannot be considered as connected by any continuous spatiotemporal process. In the strong RWR view, no concept of time can ultimately apply to this constitution, any more than any other, such as "constitution." "Reality," in this view, is a name without a concept associated to it.

20.2 Reality Without Realism and Quantum Causality

A viable physical theory must relate to a given multiplicity of phenomena or objects, which are customarily assumed to form the reality considered by this theory. I refer to both phenomena and objects, because, as Kant realized, they are not the same even in the regular human experience and, thus, in classical mechanics, which is a refinement of this experience and deals with individual classical objects or small classical systems. However, classical objects, say, planets moving around the Sun, and our phenomenal representation of them could be treated as the same for all practical purposes. This is because our observational interference could, in principle, be neglected, allowing us to consider this behavior independently. Doing so was assumed to be possible, at least in principle, in the case of all classical physical objects, even when they were not or even could not be observed, as in the case of atoms or molecules in the kinetic theory of gases, which is a classical statistical theory. Quantum phenomena put this assumption into question. Defined by the effects of the interactions between quantum objects and measuring instruments, quantum phenomena are observable in the same way as are classical physical objects and could be treated as *classical objects*. By contrast, the "uncontrollable" (quantum) nature of these interactions precludes any observation and, in some interpretations, including the one adopted here, even an inferential reconstitution, of the independent behavior of quantum objects (Bohr 1935, pp. 697, 700; Bohr 1987, v. 2, p. 62). Nobody has ever observed, thus far, a moving electron or photon, independently, to the degree that the concept of motion ultimately applies to them or any quantum objects. It is only possible to observe traces, such as spots on photographic plates, left by their interactions with measuring instruments, traces that can be described but not predicted by classical physics, thus requiring an alternative theory, such as QM or QFT, for predicting them.

I shall now introduce a concept of reality that permits the type of revision Bohr had in mind. This concept itself, however, is very general and is in accord with most, even if not all (which would be impossible), available concepts of reality in realism and nonrealism, which would, respectively, assume this reality to be representable or at least conceivable and to be beyond representation or even conception. By *reality* I refer to that which is assumed to exist, without making any claim concerning the *nature* of this existence, which thus may be placed beyond representation or even conception. I understand existence as a capacity to have effects on the world with which we interact and that, because it exists, has such effects upon itself.

In physics, the primary reality considered is that of matter, including that handled by fields. The idea of matter is still a product of thought, which, however, is customarily assumed to be a product of the material processes in the brain, and thus still of matter. Matter is commonly, but not always (although exceptions are rare), assumed to exist independently, and to have existed when we did not exist and to continue to exist when we will no longer exist. This view is upheld in the RWR-type interpretations of QM, but in the absence of a representation or even conception of the character of this existence, for example, as either discrete or continuous. Discreteness only pertains to quantum phenomena, observed in measuring instruments, while continuity has no physical significance at all. It is only a feature of the formalism of QM, which, while mathematically continuous, relates to discrete phenomena by predicting the probabilities or statistics of their occurrence.

Physical theories prior to quantum theory have been realist or (the term that is equivalent for the present purposes) ontological theories, usually *representational* realist theories. Such theories aim to represent the corresponding objects and their behavior by mathematical models, assumed to idealize how nature works. Thus, classical mechanics (used in dealing with elemental individual objects and small classical systems), classical statistical mechanics (used in dealing, statistically, with large classical systems), or chaos theory (used in dealing with classical systems that exhibit a highly nonlinear behavior) are all realist theories, as concerns the ultimate reality they consider. While classical statistical mechanics does not represent the overall behavior of the systems considered because their great mechanical complexity prevents this, it assumes that the individual constituents of these systems are represented by classical mechanics. The relativistic law of addition of velocities (defined by the Lorentz transformation) in special relativity, $s = \frac{v+u}{1+(vu/c)^2}$, for collinear motion ($c$ is the speed of light in a vacuum), runs contrary to any intuitive (geometrical) representation of motion that we can have. Relativity was the first physical theory that defeated our ability to form a phenomenal conception of an elementary physical process, a radical change in the history of physics. It is, however, a realist theory, although it complicates the nature of causality. Ultimately, photons are quantum objects and are treated by quantum electrodynamics (QED), which, in the RWR-type view does not represent the behavior of quantum objects any more than QM does.

One could define another type of realism, which is not representational. It encompasses theories that would presuppose an independent structure of reality governing the behavior of the ultimate objects considered, while allowing that this architecture cannot be represented, even ideally, either at a given moment in history or ever, but only due to epistemological limitations. In the first eventuality, a theory that is merely predictive may be accepted for lack of a realist alternative, but with a hope that a future theory will do better as a representational realist theory. Einstein adopted this view in the case of QM.

The assumption of realism of either type is precluded in the RWR-type interpretations of quantum phenomena and QM. The mathematical formalism of QM is strictly probabilistically or statistically predictive, rather than deterministic, even in considering elementary individual quantum objects and processes, which form the quantum-level reality, while suspending or even precluding a representation and possibly a conception of this reality, and an assumption that this reality is classically causal. By classical causality I, again, refer, ontologically, to the conception that the state of the system considered is determined at all future moments of time, once it is determined at a given moment of time, and by determinism, epistemologically, to the possibility of predicting the outcomes of such processes ideally exactly. As will be seen, causality may be defined differently, first, in a relativistic or local sense and, second, in a quantum-theoretical probabilistic sense. The probabilistic or statistical character of quantum predictions must be maintained by realist interpretations of QM or alternative theories (such as Bohmian mechanics) to accord with quantum experiments, where only probabilistic or statistical predictions are possible. This is because the repetition of identically prepared experiments in general leads to different outcomes, and unlike in classical physics, this difference cannot be diminished beyond the limit defined by Planck's constant, h, by improving our measuring instruments, as manifested in the uncertainty relations, which would remain valid even if we had perfect instruments.

The RWR-type interpretations do assume the concept of *reality*, defined above, as that which is assumed *to exist*, without, in contrast to realist theories, making any claims concerning the *character* of this existence, which is what makes this *reality* that of "reality *without* realism" (RWR) (Plotnitsky 2016; Plotnitsky and Khrennikov 2015). Such interpretations place quantum objects and processes either beyond representation, "the weak RWR view," or more radically, beyond conception, "the strong RWR view," which I adopt here. Not all interpretations in the spirit of Copenhagen go that far, in fact very few, if any, do. Thus, while there are indications that he has moved toward a strong RWR view in his ultimate interpretation, he never expressly stated the strong RWR view. The existence of quantum objects or what leads to this idealization (it is still an idealization) is inferred from the effects they have on the world we observed, specifically on experimental technology. In the RWR view, however, nothing could be said or, in the strong RWR view, even *thought*, concerning what happens between experiments. According to Heisenberg: "There is no description of what happens to the system between the initial observation and the next measurement. ... The demand to 'describe what happens' in the quantum-theoretical process between two successive observations is a contradiction in adjecto, since the word 'describe' refers to the use of classical concepts, while these concepts cannot be applied in the space between the observations; they can only be applied at the points of observation" (Heisenberg 1962, pp. 47, 145).

RWR-type interpretations make the absence of classical causality nearly automatic. This absence is stricly automatic if one adopts the strong RWR view, which places the ultimate nature of reality beyond conception, because the assumption that this nature is classically causal would imply at least a partial conception of this reality. However, even if one adopts the weak RWR view, which only precludes a representation of this reality, classical causality is still difficult to maintain in considering quantum phenomena. This is because to do so requires a degree of representation, analogous to that found in classical physics, that appears to be prevented by the uncertainty relations. Schrödinger expressed this difficulty in his cat-paradox paper: "if a classical state does not exist at any moment, it can hardly change causally," where a classical state is defined by the (ideally) exact position and momentum of an object at any moment of time (Schrödinger 1935, p. 154).

The question of causality is, however, a subtle matter, given that one can define concepts of causality that are not classical, and it merits a further discussion. First, I shall comment on the concepts of indeterminacy, randomness, chance, and probability, in order to avoid misundertanding concerning how these concepts are defined here, because they can be defined otherwise. In the present definition, indeterminacy or chance is a more general category, while randomness will refer to a most radical form of indeterminacy, when even a probability cannot be assigned to a possible future event. Indeterminacy and chance may be understood differently, too. These differences are, however, not germane here, and I shall for convenience only refer to indeterminacy. An indeterminate, including random, event may or may not result from some underlying classically causal processes, whether this process is accessible to us or not. The first eventuality defines classical indeterminacy or randomness, conceived as ultimately underlain by a hidden classically causal architecture; the second defines the irreducible indeterminacy and randomness. The ontological validity of the second concept of indeterminacy or randomness cannot be guaranteed: it is impossible to ascertain that an apparently indeterminate or random sequence is in fact indeterminate or random. This concept is an assumption that may only be practically justified insofar as an effective theory or interpretation based on it is developed.

I would like to add two comments on the role of probability and statistics in quantum theory from the RWR-type perspective. These remarks cannot do justice to the subject, extensively considered in literature (e.g., Khrennikov 2009; Háyek 2014). First, probability has a special temporal structure by virtue of its, correlatively, irreducibly futural and discrete character, because one can only verifiably estimate future discrete events. While true in general, this is also strictly in accord with the ultimate character of all quantum events, which, in the RWR-type interpretations, preclude us from continuously and, especially, classically causally connecting them, and about which only probabilistic or statistical predictions are possible, again, even in dealing with elementary individual events. QM or QFT is only about estimating outcomes of discrete future events, with nothing to say about what happens between these events.

The second aspect of probability that I would like to note is as follows. Randomness or indeterminacy introduces an element of chaos into order and reveals that the world confronts us with this element, even if the ultimate constitution of nature is assumed to be classically causal. It is also possible to assume the random ontology, ultimately underlying this order, thus made an illusory product of thought, an assumption that, while uncommon, has been around since the pre-Socratics. Yet another ontology contemplated by the ancient Greeks was that of the interplay of chance and necessity, which was introduced, as the atomist ontology of nature, by Democritus and developed by Epicurus and Lucretius (Lucretius 2009). However, classically causal ontology has been and remains dominant. In any event, probability brings a degree of order to our encounters with randomness or indeterminacy. It follows that probability or statistics is, too, about the interplay of indeterminacy or randomness and order, but in our interactions with the world. This interplay takes on a unique significance in quantum physics, because of the existence of quantum correlations, such as the EPR or (as they are also known) EPR-Bell correlations, found in the experiments of Einstein-Podolsky-Rosen (EPR) type and considered, in the case of discrete variables, in Bell's and the Kochen-Specker theorems, and related findings. These correlations are a form of statistical order. They are properly predicted by QM, which is, thus, as much about order as about indeterminacy or randomness, and, most crucially, about their unique combination in quantum physics. Indeed that, in certain circumstances, indeterminate or random individual events form statistically correlated and thus ordered multiplicities is one of the greatest mysteries of quantum physics.

I return to the question of causality to consider two alternatives to classical causality. Thus, the term "causality" is often used in accordance with the requirements of special relativity, which restricts (classical) causes to those occurring in the backward (past) light cone of the event that is seen as an effect of this cause, while no event can be a cause of any event outside the forward (future) light cone of that event. No physical causes can propagate, from the present to the future, faster that the speed of light in a vacuum, c, which requirement also implies temporal locality. Technically, this requirement only *restricts* classical causality by a relativistic antecedence postulate (a temporal locality), and relativity theory itself, special or general, is (locally) a classically causal and indeed deterministic theory otherwise.

Relativistic causality is, thus, a manifestation of a more general concept or principle, that of locality. This principle states that no instantaneous transmission of physical influences between spatially separated physical systems ("action at a distance") is allowed or that physical systems can only be physically influenced by their immediate environment. Nonlocality, spatial or temporal, is usually (there are exceptions) seen as undesirable. As Bohr argued in his reply to EPR, standard QM avoids it, at least in the nonrealist, RWR-type, interpretations of the theory and quantum phenomena themselves, even though under certain circumstances, such as those of the EPR-type experiments, QM can make *predictions* concerning the state of spatially separated systems, while, which was crucial to Bohr's argument, the physical circumstances of making these predictions and verifying them are local (Einstein et al. 1935; Bohr 1935; Plotnitsky 2016, pp. 136–154). The question of the locality of QM or quantum phenomena is, however, a matter of much debate and controversy, especially in the wake of the Bell and Kochen-Specker theorems and related findings, although this question was at stake in the Bohr-Einstein debate from its inception in the late 1920s. These debates cannot be addressed within the scope of this article, and the literature dealing with these subjects is nearly as immense as that on interpretations of QM (e.g., Bell 2004; Cushing and McMullin 1989; Ellis and Amati 2000; Brunner et al. 2014).

Finally, I propose the concept of quantum causality. I shall do so via Bohr's concept of complementarity, which Bohr saw as a generalization of causality. Complementarity is defined by

(a) a mutual exclusivity of certain phenomena, entities, or conceptions; and yet
(b) the possibility of considering each one of them separately at any given point; and
(c) the necessity of considering all of them at different moments for a comprehensive account of the totality of phenomena that one must consider in quantum physics.

As a quantum-theoretical concept underlain by an RWR-type interpretation, as it is in Bohr, complementarity may be seen as a reflection of the fact that, in a radical departure from classical physics or relativity, the behavior of quantum objects of the same type, say, electrons, is not governed, individually or collectively, by the same physical law, in all contexts, and specifically in complementary contexts. On the other hand, the mathematical formalism of QM offers correct probabilistic *or* statistical predictions (no other predictions are possible) *in all contexts*.

It is this probabilistic or statistical determination of what can happen as a result of our conscious decision concerning which experiment to perform at a given moment in time, that defines what I call "quantum causality." Whatever is registered as a quantum event defines a possible set of, probabilistically or statistically, predictable future events, outcomes of possible future experiments. This definition is in accord with recent views causality in quantum information theory (e.g., Brukner 2014; D'Ariano et al. 2017; Hardy 2011), except that it brings into consideration our conscious decision concerning experiments we perform, which is rarely considered. It is, however, the role of this decision that brings complementarity into play, because any such decision irrevocably rules out the possibility of our predictions concerning other, complementary, events.

One can now understand Bohr's view of complementarity as a generalization of causality (Bohr 1987, v. 2, p. 41). On the one hand, "our freedom of handling the measuring instruments, characteristic of the very idea of experiment" in all physics, our "free choice" concerning what kind of experiment we want to perform is essential to complementarity (Bohr 1935, p. 699). On the other hand, as against classical physics or relativity, implementing our decision concerning what we want to do will allow us to make only certain types of predictions and will exclude the possibility of certain other, *complementary*, types of predictions. Complementarity defines which reality can and cannot be assigned a probability to be brought about by our decision concerning what experiment to perform.

20.3 Quantum Temporality

Complementarity and quantum correlations brought with them both the role of consciousness and the futural temporality in defining our predictions concerning quantum events, and by quantum causality, the course of quantum events themselves. That does not of course mean that the concept of past is lost in dealing with quantum phenomena: the past is defined by measurements performed in the past, just as the present is defined by the measurement performed at a given present ("now") moment and possible future moments are defined by possible future measurements, all pertaining to the corresponding quantum phenomena, at which level classical concepts (space, time, motion, and so forth) and classical physics apply. They are part of the spatial-temporal continuum of our experience, where we can also use clocks (or rods) in the way we use them in classical physics or relativity, keeping in mind that relativity changes the behavior of rods and clocks depending on their local frame of reference, precluding the possibility of the universal clock. Where these concepts, including time, do not apply, in RWR-type interpretations, is to quantum objects and their behavior, and thus to how quantum phenomena come about or are related, as they are by discreteness, complementarity, or quantum correlations, which are thus beyond representation or even beyond conception. The irreducible discreteness of quantum phenomena, as is, the impossibility of assuming a continuous and especially causal process of linking them, places time "out of joint," technically, vis-à-vis any concept of time, but in any event time that can be measured by clocks in the way it is in classical physics or relativity and registered by our consciousness as such. What allows us to "set it right" is quantum causality, guided by our conscious decision what experiment to performed and enabled by the capacity of QM or QFT to give us correct probabilities or statistics of the outcomes of future experiments as determined by this decision.

At the same time, that quantum theory only predicts future phenomena or events rather than traces any past events, determined only by measurements, suggests local "time-arrows," perhaps also indicating a global "time arrow" for the available world-manifold, say, the observable universe, assuming the latter is quantum in its ultimate constitution. I shall, however, only mention this as a possibility because this constitution is currently unknown: we do not have microscopic gravity theory, which and thus whatever reconciles general relativity and quantum theory may not be quantum. I shall limit myself to local time arrows found in quantum physics. If one adopts the RWR view, one can only speak of time and time arrows as effects manifested in quantum phenomena and thus treatable objectively, and not at the level of quantum objects and behavior, which is the efficacy of these effects. There the concept of time cannot be applied any more than any other concept, such as space or motion.

One could speak here of what "time-unthinkable," defined as that in the unthinkable reality of quantum objects and behavior which governs this *discrete* registered temporality of observed events, while no continuous process connecting these events could be assumed. "Time-unthinkable" refers to those strata of the unthinkable, which are responsible for temporal effects, to the degree that one can demarcate the efficacy of these effects. This degree is limited because one can never be completely certain what ultimately contributed to this efficacy. Nevertheless, something like "time-unthinkable," also thought of as the reality underlying time, may be assumed in considering temporal effects, and for the view of time such accounts may require, in physics, philosophy, psychology, or elsewhere. The time-unthinkable is *real*, and in the present view, it belongs to matter (which is ultimately beyond conception as well), while time ultimately belongs to thought, even only to conscious thought, from which, moreover, the time used in physics, the time defined by clocks, is constructed by idealization, as Einstein clearly recognized. The ultimate "temporality," if the term applies, of the unconscious may be time-unthinkable, analogous to that found in quantum physics, to which in fact Freud refers for a parallel (on Kantian lines), in considering the unconscious, the German term for which is in fact *das Unbewusste*, the unknowable (Freud 2015, p. 121). The unthinkable? Quite possibly, even if not for Freud himself! According to J. Derrida:

> *Now* B would [in Husserl's scheme] be as such constituted by the retention of *Now* A and the protention of *Now* C; in spite of all the play that would follow from it, from the fact that each of the three *Now-s* already reproduces the same [triple] structure in itself, this model of successivity would prohibit a *Now* X from taking the place of *Now* A, for example, and would prohibit that, by a delay that is *inadmissible to consciousness*, an experience be determined, in its very present, by a present which would not have preceded it immediately but would be considerable "anterior" to it. It is the problem of the deferred effects (*Nachträglichkeit*) of which Freud speaks. The temporality to which he refers cannot be that which leads itself to a phenomenology of consciousness or of presence and one may indeed wonder by what right all that is in question here [in the efficacy of time, presence, now-ness, etc.] should still called time, now, anterior present, delay, etc. (Derrida 2016, p. 67)

The time of consciousness, conceptualized by Husserl's phenomenology as a linear sequence of presences, may be inhabited and inhibited, from the unconscious, by the efficacy that may well be the time-unthinkable, akin to that found in the efficacy (quantum objects and behavior) of quantum phenomena. Husserl's model of temporal succession is at most an idealization of experience and is not experience itself. The time used in physics, measured and, as Einstein argued, defined by clocks, is even further idealized from this model, essentially my mathematizing Husserl's concept (e.g. Weyl 1952, pp. 7–10). This idealization has served physics well from Galileo and indeed Aristotle to Einstein and beyond, including in quantum physics. While, however, the latter uses clocks and the corresponding concept of time in dealing with quantum phenomena, it tells us that this or any conception of time cannot apply to the ultimate (quantum) reality. One may refer to this structure as "quantum temporality," which is in effect correlative to quantum causality. This view reaches beyond the Kantian difference between noumena or things in themselves and phenomena, which are representations in our thought. While things in themselves are beyond knowledge, they are not beyond conception, even though the truth of this conception cannot be guaranteed (Kant 1997, p.115). In the strong RWR view, the unthinkable, including the time-unthinkable, is beyond thought.

Consider, the claim that the (known) Universe is (about) 13.8 billion years old, a time arrow extending from the Big Bang. Is this claim objective, and in what sense? Yes, but only in the following sense. There is something existing or real in the material constitution of the universe that in the interaction with our measuring instruments, beginning with our bodies and brains, in this case especially clocks, allows us to make this claim, as objectively communicable and verifiable. But what is it, this real, that is responsible for the possibility of doing is, in the RWR view, beyond our capacity to know or, in the strong RWR view, even to conceive of it, at least if this ultimate constitution of the Universe is quantum, as it appears to be, as things stand now (Plotnitsky 2016, pp. 184–186). Even apart from assuming its quantum nature, however, such concepts as space and time, as defined by rods and clocks, or otherwise, are human, derived from our experience defined by our evolutionary biological and neurological nature. They do not belong to the constitution of the Universe. Quantum physics, however, radically prevents us from using them in considering quantum objects and behavior, while allowing us to use them in considering quantum phenomena, which applicability (as objective as in classical physics or relativity) enable us to do quantum physics as experimental-mathematical science of nature.

According to S. Coleman, "if thousands of philosophers spent thousands of years searching for the strangest possible thing, they would never find anything as weird as quantum mechanics" (cited in Randall 2005, p. 117). That may be true! But, defined by our interactions with nature, by means of our thought, brain, and technology, even this weirdness is a product of our thought, thus connecting thought and the quantum. Perhaps poets can do better. Shakespeare, who gave us Hamlet's "time our joint," with which I started here, may be given the last word, now with Hamlet's remark to Horatio:

> There are more things in heaven and earth, Horatio,
> Than are dreamt of in your philosophy.
> *The Tragedy of Hamlet, Prince of Denmark* (Act I.4, 165–166)

Some editions actually say "our philosophy"; "your philosophy" makes Hamlet more suspicious of philosophy, makes him more akin to a physicist perhaps. Physics helps us to discover these things and help philosophy to understand them, and keeps its honesty, as Nietzsche said. "And therefore, long live physics!" And then, *Hamlet*, a play, which question realism and causality as much as any work of literature or philosophy, takes place in Denmark, too, where this challenge in physics defined *der Kopenhagener Geist der Quantenheorie*.

Acknowledgments I would like to thank Acacio de Barros, Mauro D'Ariano, Ehtibar Dzhafarov, Gregg Jaeger, Andrei Khrennikov, and Paolo Perinotti for valuable discussions concerning the subjects addressed in this article.

References

Bell, J. S. (2004). *Speakable and unspeakable in quantum mechanics*. Cambridge: Cambridge University Press.

Bohr, N. (1935). Can quantum-mechanical description of physical reality be considered complete? *Physical Review, 48*, 696–702.

Bohr, N. (1987). *The philosophical writings of Niels Bohr*, 3 vols. Woodbridge: Ox Bow Press.

Brukner, C. (2014). Quantum causality. *Nature Physics, 10*, 259–263.

Brunner, N., Gühne, O., & Huber, M. (Eds.). (2014). Special issue on 50 years of Bell's theorem. *Journal of Physics A, 42*, 424024.

Cushing, J. T., & McMullin, E. (Eds.). (1989). *Philosophical consequences of quantum theory: Reflections on Bell's theorem*. Notre Dame: Notre Dame University Press.

D'Ariano, G. M., Chiribella, G., & Perinotti, P. (2017). *Quantum theory from first principles: An informational approach*. Cambridge: Cambridge University Press.

Derrida, J. (2016). *Of grammatology* (trans. Spivak G. C.). Baltimore: Johns Hopkins University Press.

Einstein, A., Podolsky, B., & Rosen, N. (1935). Can quantum-mechanical description of physical reality be considered complete? In J. A. Wheeler & W. H. Zurek (Eds.), *Quantum theory and measurement* 448 (pp. 138–141). Princeton NJ, 1983: Princeton University Press.

Ellis, J., & Amati, D. (Eds.). (2000). *Quantum reflections*. Cambridge: Cambridge University Press.

Freud, S. (2015). The unconscious. In S. Freud (Ed.), *General psychological theory* (pp. 116–150). New York: Collier, 1963.

Hardy, L. (2011). Foliable operational structures for general probabilistic theory. In H. Halvorson (Ed.), *Deep beauty: Understanding the quantum world through mathematical innovation* (pp. 409–442). Cambridge: Cambridge University Press.

Háyek, A. (2014). *Interpretation of probability, Stanford encyclopedia of philosophy* (Winter 2014 ed.), E. N. Zalta (Ed.). http://plato.stanford.edu/archives/win2012/entries/probability-interpret/

Heisenberg, W. (1930). *The physical principles of the quantum theory* (trans. Eckhart, K., Hoyt, F. C.). New York: Dover, Reprint 1949.

Heisenberg, W. (1962). *Physics and philosophy: The revolution in modern science*. New York: Harper & Row.

Kant, I. (1997). *Critique of pure reason* (trans. Guyer, P., Wood, A. D.) Cambridge: Cambridge University Press.

Khrennikov, A. (2009). *Interpretations of probability*. Berlin: De Gruyter.

Lucretius, T. C. (2009). *On the nature of the Universe* (trans. Melville, R.), Oxford: Oxford University Press.

Plotnitsky, A. (2012). *Bohr and complementarity: An introduction*. New York: Springer.

Plotnitsky, A. (2016). *The principles of quantum theory, from Planck's quanta to the Higgs boson: The nature of quantum reality and the spirit of Copenhagen*. New York: Springer.

Plotnitsky, A., & Khrennikov, A. (2015). Reality without realism: On the ontological and epistemological architecture of quantum mechanics. *Foundations of Physics, 25*(10), 1269–1300.

Randall, L. (2005). *Warped passages: Unraveling the mysteries of the universe's hidden dimensions*. New York: Harpers Collins.

Schrödinger, E. (1935). The present situation in quantum mechanics. In J. A. Wheeler & W. H. Zurek (Eds.), *Quantum theory and measurement* (Vol. 1983, pp. 152–167). Princeton: Princeton University Press.

Weyl, H. (1952). *Space time matter* (trans. Brose, H. L.). Mineola: Dover.

MIX
Papier aus verantwortungsvollen Quellen
Paper from responsible sources
FSC® C105338

If you have any concerns about our products,
you can contact us on
ProductSafety@springernature.com

In case Publisher is established outside the EU,
the EU authorized representative is:
Springer Nature Customer Service Center GmbH
Europaplatz 3, 69115 Heidelberg, Germany

Printed by Libri Plureos GmbH
in Hamburg, Germany